Mars Awaits: Humanity Ventures

Sanjay

Table of Contents

Chapter 1

Introduction

The idea of exploration fulfils human curiosity in order to find answers to the science mysteries present in the Universe. The question that has been pursued over many years now is: how did life evolve in our solar system, and other than Earth where else life could have existed in the early days of our solar system formation? These questions lead us to identifying the habitable zone of our solar system and the planets that fall under this habitable zone. Earth and Mars are prime candidates for satisfying the criteria defined to be in the habitable zone of the solar system. For many years now we have been sending unmanned robotic probes, orbiters, landers and rovers to the red planet Mars in order to understand its composition, atmosphere, geology etc. There have been over 45 missions that were targeted to Mars by collective effort from different space agencies from around the world. The unmanned robotic missions are great at exploring the red planet while also surviving the harsh atmosphere environment and radiation dose. The robotic exploration has paved the way for humanity to research technology that supports human settlement on Mars. The rovers Spirit and Opportunity have led a foundation for developing technology to send robust and autonomous systems to conduct experiments on Mars. The orbiters have mapped the surface of the planet with great detail that helps space agencies in selecting landing sites of future missions in order to define mission objectives, and also to have knowledge about the environment in which the lander or rover will be performing experiments. The orbiters like Mars Reconnaissance Orbiter (MRO) and Mars Express have mapped the interesting regions of the planet like Volcanoes, Canyons and polar ice caps. Even after using the best quality Hires Images from MRO, the surface has only been mapped with resolution of about 20 m per pixel. For this purpose the rovers have been designed and sent to Mars in order to partially fill the data gaps from the orbiters. The rovers have been successful in doing onsite research about mineral composition of rocks and soil, atmospheric studies along with sending beautiful panoramic mosaics of the red planet. The rovers are still limited by their ability to move quickly from one site to another site of interest in order to conduct science experiments.

1.1 Motivation

The limitation posed by ground based exploration vehicles are a constraint to going further on the red planet. In order to overcome this limitation, the idea of aerial exploration has been considered for many years and many methods have been proposed to go a step further by changing the mode of exploration. Aerial exploration is great at moving from point A to point B without considerations about the terrain and ground obstacles. Aerial exploration also allows for movement at higher speeds from one site to another site of interest. There have been many ideas and theories that propose modes of transport that consider flying or hovering v ehicles. Among many projects considered for developing multi rotor systems to fly on Mars, the project that has been an inspiration to this book is the Ingenuity Mars helicopter project. [1] The Mars helicopter is a small, lightweight helicopter design that has been designed and tested by JPL in NASA in a simulated environment that represent atmosphere, density, pressure and gravity in Mars-like conditions. This master book is based on two aspects of designing a multi rotor Unmanned Aerial Vehicle for Mars exploration. The first aspect of the project will focus on studying the thin atmosphere aerodynamics on Mars and proposing a blade profile for a Mars UAV and also designing a CAD model of the UAV for CFD simulations. The other aspect of the book will focus on modelling the equations of motion for an octa-rotor coaxial UAV, designing a PID controller and simulating the model in SIMULINK to perform some basic autonomous task like hovering, following a trajectory etc. The technological demonstration about making the first powered fl ight in the Mars at mosphere from the Ingenuity helicopter will build a foundation for developing autonomous control for rotor crafts for space purposes. There have been a significant amount of research done by JPL and Caltech in order to develop thin airfoil aerodynamics analysis that focuses on low Reynolds number flow. A significant part of these studies have been conducted for the Ingenuity helicopter p roject. This book project will take inspiration from the Ingenuity helicopter and develop a multi rotor system that will be able to make an autonomous flight in the Mars atmosphere.

1.2   UAV Flight in Mars Atmosphere

Mars has a very challenging atmosphere in terms of temperature fluctuations, dust storms, surface composition that limits the locations that Mars exploration rovers can reach. The concept of UAVs to fly on Mars is proposed to prove that with considerable optimization in rotor blade design, enough lift can be generated to fly a lightweight UAV in the thin atmosphere. The concept also focuses on making the flight and operation autonomous, mapping the surrounding terrain, and path planning to assist the ground based rover to reach beyond its current capabilities. The Mars UAV will operate in the conditions where the tip Mach numbers are high and Reynolds number is low. In a generic rotor design, it is important to maintain subsonic speed at the tip of the rotor in order to avoid undesired shock wave formation. The generated shock waves highly affect the lift generation capabilities of the rover if not considered beforehand. Because the atmospheric density is very low on Mars, there is a benefit in spinning the rotor at higher RPM and still keep the tip speeds subsonic. The hovering of the vehicle will be controlled in a similar manner to any quadrotor that flies in Earth conditions. The roll, pitch and yaw movement commands will be handled by the proposed PID controller that will be designed specially to control the co-axial rotors. The reduced value of gravity will help the vehicle remain stable when flying and will minimise small instabilities caused by the frequencies of unstable phugoids [2]. The proposed size of the rotor blade is 1.12 meters and two rotors spin in opposite directions when mounted in a co-axial manner. This approximation estimate the overall UAV mass to be around 6 Kilograms. Parts of the on board payload and system requirements will be discussed in the CAD modelling part of this project. Currently the Mars rovers that are operating on Mars are powered by RTG (Radioisotope Thermo electric Generator), But because the RTGs have very low efficiency, they are not suitable for UAV because of the heavy subsystem required to manage the heat produced by RTG. The Mars UAv is designed to get its power from solar energy. The larger arms of the Mars UAV are suitable to mount roll out solar arrays on them. These solar arrays can be extended for charging and can be retracted during flying. Flight data from the Ingenuity helicopter project will clarify whether or not a powered fight is possible in Mars atmosphere, and will help this idea to be pursued in terms of increasing the payload mass and decreasing system mass [1].

1.3   book **Outline**

1.3.1   Chapter 2

This chapter of the master book will give insights to the basics of computational fluid dynamics and Navier-Stokes equations. The goal of the literature review on reynolds averaged Navier-Stokes equations is to adapt the understanding of RANS for low reynolds number flow simulations. This chapter also talks about the generic and blade element momentum theory of rotors.

1.3.2   Chapter 3

In this chapter the low reynolds number aerodynamics in thin atmosphere has been studied. The flow performances of flat and cambered plates has been studied in order to optimize for delayed transition from laminar to turbulent flow. This section of the thesis also talks about the Mars Helicopter project that has been developed by JPL to make a helicopter model for a tech demo on UAV flight in Mars atmosphere.

1.3.3   Chapter 4

This chapter of the book project shows the simulation results for two dimensional flow around flat plate, cambered plate and Ingenuity helicopter rotor blade airfoil. In this chapter the studies from different authors is compared for optimizing a 2D airfoil for low reynolds number flow. The effects of high mach number around the rotor tips is also described in this chapter.

1.3.4   Chapter 5

In this chapter the simulation results from Ansys fluent are presented. The Mars condition simulation for flow around the single and co-axial rotor system is considered to see the effects of different RPM values and optimize for best rotor speed to produce necessary thrust.

1.3.5   Chapter 6

In this chapter a CAD model of the Mars UAV is presented. The different subsystems of the model, the vehicle structure and power system is described in this section.

1.3.6   Chapter 7

In this section of the master book, the dynamics of a UAV model is described. Conventional quadrotor is taken as reference in order to describe the mathematical model of the Mars UAV.

1.3.7   Chapter 8

This section of the book introduces control strategies for the Mars UAV. Open loop and closed loop simulations were performed and the model was extended to implement a PID controller to record the response of the Mars UAV model for a trajectory following task.

1.3.8   Chapter 9

The final chapter of the master thesis considers concluding remarks from the work done on the design and control of Mars UAV. Learning outcomes and future control implementations are also suggested for the Mars UAV.

Chapter 2

Literature review and Background

In order to explain the work done in this book on the flow simulations part, it is necessary to introduce the basics of fluid dynamics, CFD, Quadrotor equations of motions etc. The concept of a Mars UAV is proposed to operate in incompressible flow conditions and for that the used Navier-Stokes equations are from the incompressible set of equations. For numerical simulations and model building, the Fluent from ANSYS software is used. The 2D analysis of the different airfoils are performed using the Xfoil tool.

2.1   Computational Fluid Dynamics

Computational Fluid Dynamics is based on the Navier-Stokes equations that are used to model numerical solutions to describe flow behaviours. These are also key equations to understand flow interactions around solid surfaces. The numerical model derived from Navier-Stokes equations is beneficial in terms of doing a simulation for the Mars condition before fabricating the model for lab experiment. Because of evolving technology and modern computers, the time required to simulate the flow has been greatly reduced and complex experiments can be modeled inside a CFD environment before the actual experiments.

Using the three fundamental laws of physics, which are conservation of mass, conservation of energy and newton's second law of motion; the Navier-stokes equations create set of partial differentiation equations. In order to solve for primary flow variables that are pressure and velocity, the conservation of mass and conservation of momentum are sufficient. The energy conservation is necessary when the compressible flows are discussed. Since this thesis is only focusing on the incompressible flows, we will not discuss the energy equations. The law of conservation of mass states that mass can not be created not destroyed. So for the incompressible flow, the net flow entering the boundary volume is equal to the net flow leaving the volume. The boundary volume is a box shaped enclosure defined in the CFD set up. The dimensions of the volume are $3400 \times 1500 \times 1500 \; mm^3$

$$\frac{\partial \rho}{\partial t} + \frac{\partial (\rho u_i)}{\partial x_i} = 0 \tag{2.1}$$

The equation (2.1) is called the continuity equation in aerodynamics. The flow considered is the incompressible flow thus the density does not change over time.

$$\frac{\partial \rho}{\partial t} = 0 \tag{2.2}$$

Therefore,

$$\frac{\partial (\rho u_i)}{\partial x_i} = 0 \tag{2.3}$$

Newton's second law of motion says that the total force is equal to the change in momentum over time. The conservation of momentum for the defined volume is defined by the equation (2.4) in Cartesian coordinates frame.

$$\frac{\partial (\rho u_i)}{\partial t} + \frac{\partial [\rho u_i u_j]}{\partial x_j} = -\frac{\partial p}{\partial x_i} + \frac{\partial \tau_{ij}}{\partial x_j} + \rho f_i \tag{2.4}$$

Where, τ is stress, ρ is density, p is pressure u, v and w are the velocity component in x,y and x directions respectively. velocity and f is the force.

2.1.1 Reynolds Averaged Navier-Stokes equations

In fluid mechanics an instantaneous solution is not preferred in most of the areas and instead an average solution is computed from the set of equations described earlier. The purpose of RANS is also to separate the turbulent and laminar components in the flow equations. By inputting the expected values of RHS and LHS, on all four equations and simplifying according to the rules of expectation, the Reynolds Averaged Navier-Stokes equations are derived as,

$$\frac{\partial u_i}{\partial x_i} = 0 \tag{2.5}$$

$$\frac{\partial u_i}{\partial t} + u_j \frac{\partial u_i}{\partial x_j} = f_i - \frac{1}{\rho}\frac{\partial p}{\partial x_i} + \nu \frac{\partial^2 u_i}{\partial x_j \partial x_j} \tag{2.6}$$

Averaging on the right and left hand sides of the equation (2.6) we get,

$$\frac{\partial \bar{u}_i}{\partial t} + \bar{u}_j \frac{\partial \bar{u}_i}{\partial x_j} = \bar{f}_i - \frac{1}{\rho}\frac{\partial \bar{p}}{\partial x_i} + \nu \frac{\partial^2 \bar{u}_i}{\partial x_j \partial x_j} - \frac{\partial \overline{u'_i u'_j}}{\partial x_j} \tag{2.7}$$

In the equation (2.7), $\bar{u}_i$ is the averaged value of velocities in the Cartesian x, y and z directions and f_i denotes the total forces acting on the bounding volume. The new set of equations referred to as Reynolds Averages Navier-Stokes equations introduces 6 new unknown variables due to the symmetric Reynolds stress tensor [2]. Including the new set of equations, there are total 10 unknown variables are present. There are only four equations to solve for 10 unknown variables thus, In order to derive the amount of equations needed, the evolution equations of the stress tensor needs to be derived. Since this goes out of the focus of this book, the derivation of the remaining equations will not be discussed in this book.

2.2 Momentum theory of Rotors

The interest within the robotics community has peaked recently in order to understand the aerodynamics of quadrotors to improve the rotor efficiency by including the lift and drag parameters into the mathematical models. The studies in this field have only been focused on the theory of momentum for helicopters and indeed more detailed studies of rotor blade aerodynamics is necessary in order to include the aerodynamic parameters into the vehicle from a control prospective. [3] The momentum theory for rotary wing was developed in the early stage for aircraft propellers. It is considered as one of the most widely used theories in the field of propeller analysis for subsonic flights. The simple approach of the theory has made it the base foundation for modelling the aerodynamic forces on rotors. This theory considers the rotor as an actuating disk with with the accelerating fluid medium forming a stream tube in the axial direction. The visual representation of the stream tube can be seen in Figure 2.1. The working fluid (normally air) is sucked and accelerated by the rotors in the downward direction which generates a virtual induced flow of fluid which has velocity $v^i = (v^i_x, v^i_y, v^i_z)^T$ [3].

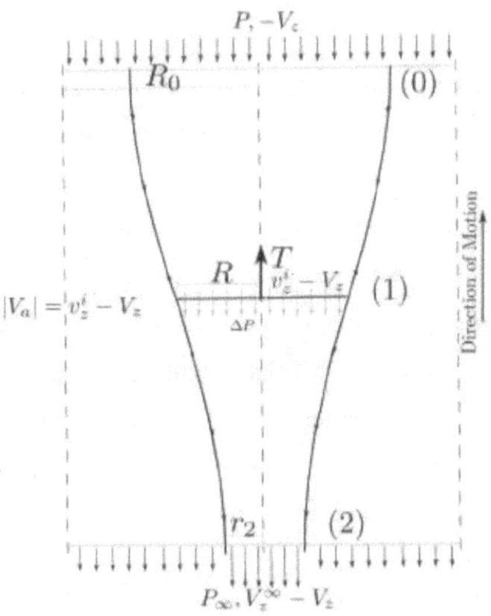

Figure 2.1: Control volume for streamtube representation

The disk loading is defined as the ratio between the Thrust and disk Area.

$$DL = \frac{T}{A} \qquad (2.8)$$

According to the assumptions made in the momentum theory, it is clear that the the higher the disk loading values the more likely the assumptions will hold [3]. For quadrotors, the observed value of disk loading is higher as compared to the helicopters and thus the momentum theory fits better for quadrotors than it does for the helicopter vehicles. The fluid flow velocities have generic form in the Cartesian system, thus it is recommended to consider a single direction in order to optimize the flow equations. The discussion is based on the flow conditions in the axial or Z direction. From the Figure 2.1(Courtesy of Bangura [3] the velocity in the axial Z direction can be defined as $V = (0,0,V_z)^T$.

The fluid flow through the rotor disc is continuous and is given by a constant speed at which the vehicle hovers. This speed is defined as the induced velocity in the Z direction or is denoted by v^i. When the rotor disc is spinning there is a pressure gradient observed on the inlet and outlet sides and this pressure gradient is denoted by ΔP. The bounding volume that is defined to contain the rotor disk has a radius of R_0 where the radius of the rotor disk is R. The bounding volume is defined with the intention to compute the amount of fluid entering from the inlet and leaving from the outlet. The upstream plane has radius R_0 and the downstream plane has the radius of $r < R$. It is obvious that is has a lesser radius on the outlet side due to the pressure gradient generated and the flow lines shrink as they pass through the rotor. The assumption made here implies the the flow stream lines are parallel with the axial Z axis or the rotor spin axis. The thrust force can be computed from the equation of momentum theory by considering the difference in mass flow per volume. This gives the following equation (2.9),

$$T = \dot{m}(V_z^\infty - V_z) - \dot{m}(-V_z), = \dot{m}V_z^\infty \tag{2.9}$$

Where $\dot{m}$ is the mass flow rate and the mass flow rate is given as

$$\dot{m} = \rho A |V^a| \tag{2.10}$$

Where,

$$V^a = \begin{pmatrix} 0 \\ 0 \\ v_z^i - V_z \end{pmatrix} \tag{2.11}$$

Because the mid section radius of the stream tube and the radius of the rotor are same, the cross section area is given by $A = \pi R^2$. The power can be calculated from the force and velocity. The force here is considered as thrust force thus,

$$P_T = T(v_z^i - V_z) \tag{2.12}$$

This power can also be considered as equivalent to the change in the kinetic energy at the inlet and outlet. Therefore the equation (2.12) can also be written as,

$$P_T = \frac{1}{2}\dot{m}(V_z^\infty - V_z)^2 - \frac{1}{2}\dot{m}(V_z)^2,$$

$$P_T = \frac{1}{2}\dot{m}(V_z^2 - 2V_zV_z^\infty + (V_z^\infty)^2) - \frac{1}{2}(V_z)^2. \tag{2.13}$$

From the equations (2.9) and the equations (2.12) the equation (2.13) can be modified and written as,

$$\dot{m}V_z^\infty V^a = \frac{1}{2}\dot{m}((V_z^\infty)^2 - 2V_zV_z^\infty)$$

Therefore,

$$v_z^i = \frac{1}{2}V_z^\infty$$

And the equations for Thrust power and Thrust force can be derived as,

$$T = 2\rho A v_z^i (v_z^i - V_z), \tag{2.14}$$

$$P_T = T(v_z^i - V_z) \tag{2.15}$$

The equations (2.14) and (2.15) are considered as the Thrust and Thrust power in the axial or Z direction of the fluid flow. The direction of rotor spinning is also considered to be the same as the

Z-axis. These equations [3] give the values in steady state and in one direction of fluid flow motion. If needed, the equations for generic motion of the rotor disk can also be derived by referring to the equations (2.14) and (2.15). The generic motion of rotor is not discussed here because it is out of scope of the research for this book.

2.2.1 Blade Element Momentum theory

In the generalised momentum theory, there have been number of assumptions made which make it less reliable to extract flow p arameters. For this purpose B lade Element theory w as proposed. The blade element theory uses the geometrical profile of the cross section of the rotor in order to compute aerodynamic forces acting on the rotor. This results in a higher number of equations to solve for the Forces and Moments. The rotor twist distribution is also taken into consideration in the blade momentum theory.

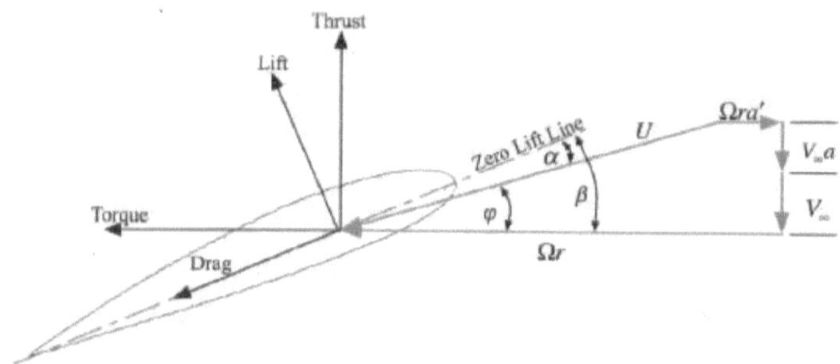

Figure 2.2: Aerodynamic forces on blade profile

In the blade element theory also a stream tube is considered to be the control volume inside which a rotor is placed. In the plane of the rotor, the boundaries of the stream tube splits the rotor into number of elements with the width given as dr collectively representing the rotor blade section. At every blade element the differential thrust can be defined as dT and can be defined for the particular cross-section of that element. The torque is given as dQ. From Figure 2.2(courtesy of Rwigema [4]) it can be seen that in order to compute the Thrust and Torque, equations need to be written in term of the angle of attack which is given by the angle ϕ. From here on, the angle of attack is defined for each element of the rotor blade which is being considered to calculate total thrust and torque. From the Figure 2.2 the incremental thrust and torque is given as equations (2.16) and (2.17).

$$dT = \frac{1}{2}B\rho U^2(C_L\cos\phi - C_D\sin\phi)cdr \tag{2.16}$$

$$dQ = \frac{1}{2}B\rho U^2(C_L\sin\phi + C_D\cos\phi)crdr \tag{2.17}$$

Aerodynamic performance of a rotor can be computed from the combination of the equations derived in the Momentum theory and the Blade Element theory. For this the blade profile or airfoil shape, twist distribution are assumed to be known. By further modifying the equations (2.16) and (2.17), the thrust and torque can be defined in terms of the known variables from the Figure 2.2.

$$dT = \sigma'\pi\rho\frac{V_\infty^2(1+a)^2}{\sin^2\phi}(C_L\cos\phi - C_D\sin\phi)rdr$$

$$dQ = \sigma'\pi\rho\frac{V_\infty^2(1+a)^2}{\sin^2\phi}(C_L\sin\phi + C_D\cos\phi)r^2dr$$

In this book, aerodynamic analysis plays a crucial role in quadrotor performance. The BEMT (Blade Element Momentum Theory) gives accurate values of aerodynamic coefficients in order to model them into the numerical analysis to get the rotor performance. On Mars the flow is assumed to be incompressible and the analysis is done for the low Reynolds number flow, additionally the aerodynamic coefficients are affected by the viscosity e ffects. This is also part of the reason for considering the Viscous model in Fluent while simulating the rotor blade in ANSYS. In this book, XFOIL tool is used to consider the viscosity effects. In XFOIL it is possible to input the Reynolds number and tip mach number as input in order to get the viscosity information of the flow. Reynolds number define the property of the flow in terms of streamline behaviour of the flow and the Mach number gives the ratio of velocity to the speed of sound in the considered medium of analysis.

$$Re = \frac{\rho V_\infty c}{\mu} \tag{2.18}$$

$$M = \frac{V_\infty}{\sqrt{\gamma RT}} \tag{2.19}$$

Chapter 3

Rotor blade optimization for Mars

This chapter of the book will briefly talk about the importance of blade profile optimization for the low Reynolds number and high mach number flow in Martian atmospheric conditions. Mars surface pressure is very low as compared to Earth, thus the blade profiles in the directory of NACA series is not as efficient i n p roviding t he r equired l ift a s t hey a re d esigned t o o perate i n Earth conditions.

(a) Subsonic NACA 0012 airfoil   (b) High Mach number CLF 5605 Airfoil

Figure 3.1: subsonic and transonic airfoils

The comparison between the subsonic and transonic airfoils are shown in Figure 3.1. The NACA series airfoils are widely used in rotor designs for multi rotor crafts. The twist distribution also plays an important role in providing correct geometry for optimized aerodynamic coefficients. The atmospheric conditions of considered Mars and Earth atmosphere are mentioned in the table 3.1.

Parameter	MARS	EARTH
Density (Kg/m^3)	0.00155	1.23
Pressure (Pa)	636	101325
Temperature (K)	214	288
Speed of Sound (m/s)	230	340
Gamma, γ	1.3	1.4

Table 3.1: Mean Atmospheric Properties

The lower value of lift generated by a rotor in Mars atmosphere is partially compensated by the low gravity of the planet. This means that there is a slight advantage to the rotor craft flying in Mars atmosphere because the gravitational acceleration is low compared to Earth. The drawback is the temperature and atmospheric composition of Mars. The average surface temperature on Mars is about 214 K and the atmosphere is mostly (95%) composed of CO_2. This results in lower speed of sound based on equation 2.19. The challenge of lower speed of sound constrains the maximum tip speed that the rotor can achieve because the lower the speed of sound, the sooner the flow becomes supersonic. As soon as the flow crosses the subsonic limit, the undesired shock waves are generated. These shock waves drastically increase the leading edge drag force and decreases the overall aerodynamic lift-generating capabilities of the rotor. Prior research in airfoil optimization and 2D flow analysis is needed in order to further advance the existing performance values of blade profiles in low Reynolds number flow conditions.[5][6]

3.1   Low Reynolds Number Aerodynamics

The need for considering the low Reynolds number aerodynamics for the Mars UAV is discussed in the previous sections. When dealing with the low Reynolds number flow, the boundary layer is assumed to be fully laminar up to the point at which it separates from the rotor surface. The subsequent turbulent flows are not observed until the point of separation this is mainly because the separation point is delayed further towards the trailing edge when low reynolds number and low angle of attack are considered. In Figure 3.2, the laminar flow separation points are seen for a range of Reynolds number as well as with the change in angle of attack.

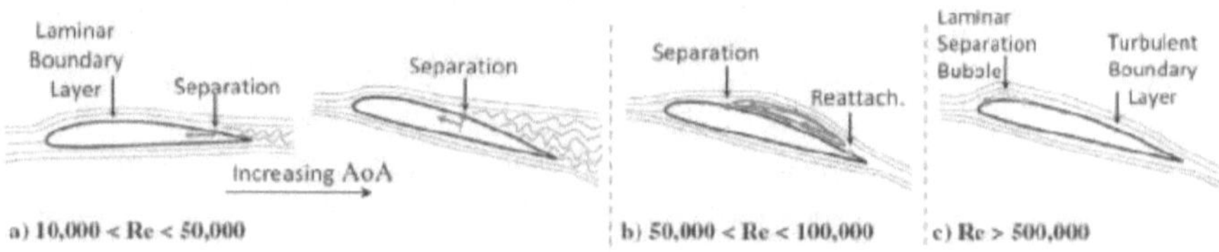

Figure 3.2: Flow separation affected by Reynolds number and angle of attack

At this stage it is important to understand the critical Reynolds number. The critical Reynolds number is defined as the Reynolds number at which the laminar flow is transitioning towards the turbulent behaviour. The super critical Reynolds number is defined as the the range of Reynolds number where the turbulent transition is always occurring before the laminar flow separation. Because the understanding of compressible flow at low Reynolds number is limited, the detailed study is not usually performed because it is not coherent with the scientific goal of doing aerodynamic analysis for Mars atmosphere conditions. This is mainly because the flow considered for the Mars UAV is incompressible. The generic series of NACA airfoils are sensitive to change in Mach number where as the flat plate and cambered airfoils flow is not heavily mach number dependant.[5][6] The region of transition between the sub critical and super critical Reynolds number flow is challenging because of the external disturbances, vibrations and the anomalies on the smoothness of the surface.[7][8][9] The reason for flat plates and cambered airfoils are not highly dependent on the Reynolds number is observed to be because of the sharp leading edge of the airfoils compared to the conventional airfoils.[7] Even if the cambered plates are not dependent on the flow Reynolds number, it is difficult to carry out analysis with conventional methods due to the leading edge flow separation on these airfoils. In the modelling when the flow Reynolds number is significantly decreased, the separation layer has the ability to reattach and this is because the sharp leading edge not allowing the generated shear layer to transition from laminar to turbulent. In the previous research, this laminar streamline re-attachments is seen for the flat plate aerodynamics for the Reynolds number range of approximately 10,000 [10][11].

The laminar to turbulent transitions are not only limited to the leading edge of the flat plates but are also observed in the trailing edge of the airfoils. This behaviour is referred as vortex shedding and is normally observed near sharp edges or surfaces with uneven geometry. The vortices are formed due to the instabilities present in the flow and are also observed at the flow separation regime of the airfoils. The pressure gradient also plays an important role here in the formation of the shedding vortices. The laminar to turbulent transition is considered as minor anomaly as compared to the significance of the shedding vortices formed in the low Reynolds number analysis[12].

3.1.1 Performance of Flat and Cambered Plate Airfoils

Flat plates that have a leading edge that is sharp have different performance than the conventional airfoils in low Reynolds number. In-depth studies on this behaviour suggest that the transitioning region from laminar to turbulent differs when flat plate and a conventional airfoil is considered. The studies from Hoerner [13] shows that the lift coefficient for a cambered plate is lower than the lift coefficient for the N60 airfoil. However the C_L is observed to decrease by a very small amount for the flat plate as compared to the N60 airfoil when the low Reynolds number is considered. The plot in Figure 3.3 and Figure 3.3 (courtesy of Koning[14] and Hoerner[13]) shows the behaviour of lift coefficient C_L with varying Reynolds number of the flow.

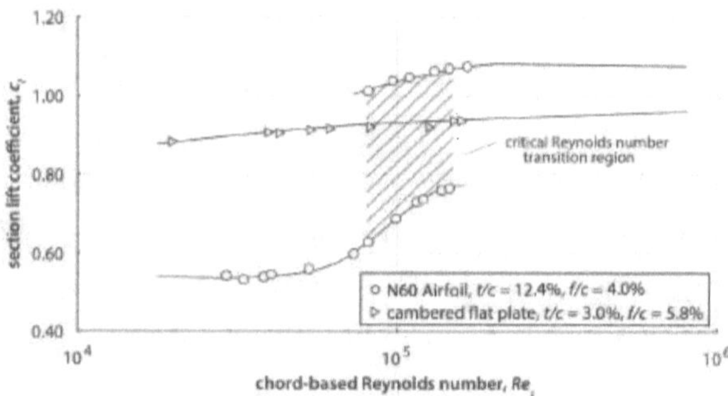

Figure 3.3: CL vs Reynolds number

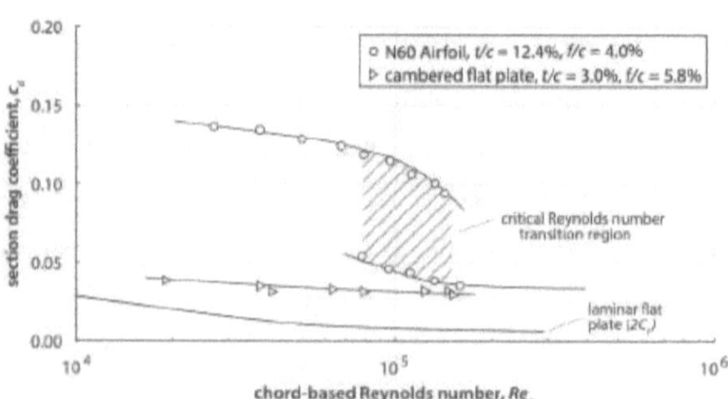

Figure 3.4: CD vs Reynolds number

The thickness to chord ratio of an airfoil represents the thickness of the airfoil at different distances from the leading edge. This ratio is a key parameter when examining the performance of airfoil at transonic speeds. This behaviour is observed at constant angle of attack and the flat plates that are considered for this analysis have the ratio of thickness to chord as 3%. The airfoil camber for the flat plate is 5.8%. These values for the considered N60 airfoil are 12.4% and 4% respectively.[14][13] From the Figure 3.3 and Figure 3.3 is is observed that the Lift coefficient and the Drag coefficient is nearly independent of change of the Reynolds number in the range of 10^4 to 10^6. The sharpness of the leading edge has an effect on the location of the separation point. The stagnation point is defined as the point on the airfoil surface at which the free stream velocity becomes zero. One stagnation point occurs at the leading edge and the other occur at the trailing edge of the airfoil. With the increase in the angle of attack, the stagnation point changes position and is moved towards the lower surface. This creates turbulent behaviour in the flow and this allows for a longer range of Reynolds number before the super critical Reynolds number is reached. When the angle of attack is fixed and is not varying, then the flow does not show the transition occurring at the super critical Reynolds number because the flow breaking point is shifted downwards. The turbulent edge of the flow always occurs at the same point for all non zero angles of attacks [7]. The turbulent transition flow characteristic is observed at $x/c = 2.5\%$ (x/c is the thickness to chord ratio) and at the Reynolds number range from 10^4 to 10^5 [15]. For flat plates the turbulent re-attachments is observed at angles of attack ϕ around 7 to 10 degrees [7]. This observations is also supported by other studies which suggest the formation of leading edge bubble is suppressed by the formation of multiple small vortices on the top surface of the flat plate. This was observed for the Reynolds number range of 10^4 and at angle of attack $\phi = 8°$.[16]

3.1.2 Laminar and Turbulent Reattachments

The hysteresis observed in the conventional airfoils does not occur in the case of thin plates and this is because the nose turbulence increases way before the pressure rises on the upper surface [7]. From the analysis it can be assumed that there has to be a range of Reynolds numbers where the flow does not become turbulent from laminar even if the sharp edge is considered. Therefore in the range of low Reynolds number, the flat plate have shown behaviour of the laminar flow in stead of directly transiting into turbulent flow [10][11].

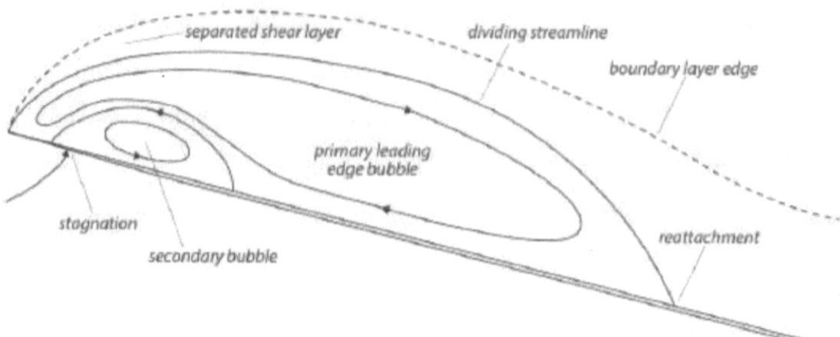

Figure 3.5: Separation bubble

Camber line is the hypothetical line drawn from the leading edge to the trailing edge while following equal distance from the top and bottom line of the airfoil. From the definition of the camber line it can be understood that the the more the camber line is bent, the less angle it makes with the incident free stream of flow. Thus, the positive effect of having a significant camber is that the angle between the camber line and the incident free stream flow line is reduced. The sharp leading generates turbulent behaviour in the flow and thus it affects the lift generation capabilities of the flat plate. In contradiction the camber of the airfoil assists in increased lift generation. However this relation only holds up to a certain angle of attack before the stalling takes place. The camber of the airfoil also assists in holding the flow stream close to the surface which aids in additional lift generating capability [7]. The different regions of flow lines attachment are shown in Figure 3.5 (courtesy of [17] and [6]. The results from experiments are contradictory in terms of understanding the turbulent reattachments of the shear layers. as shown in Figure 3.5, the laminar separation bubble is observed from the analysis of pressure distribution from the wind tunnel testings as well as the pressure sensitive paint experiments. In the laminar separation bubble the reattachments are disappearing as the Reynolds number is increasing beyond the range of 10^4 Anyoji *et al.* [18]. There is a limit to the life time of the separation bubble at the end of laminar separation bubble. Towards the trailing edge as the stream lines are transitioning towards the disturbances or turbulent behaviour, the life span of the bubble decreases. After crossing a limit of transitioning Reynolds number, the bubble bursts and this results in the vortex creation at the trailing edge. Thus, from the previous section and from the understanding of the laminar separation bubble it is clear that the aerodynamic performances of a rotor blade designed for Mars atmosphere has a very thin line of margin before the flow becomes turbulent as the tip Mach number increases. Therefore the flight of the rotor crafts designed for mars atmosphere occur at a specific combination of the tip mach number and the Reynolds number of the flow.

The standard two dimensional flow analysis assumes a 2D boundary layer around the airfoil. Even if the assumed boundary layer is thin, the comparison between the thin layer boundary layer, RANS (Reynolds Averaged Navier Stokes equations) show only small changes in the coordinates along the camber line at which the laminar separation bubble occur.[6]

3.1.3   Ingenuity Helicopter

As discussed in the previous sections, the Ingenuity helicopter is designed to fly as a part of payload aboard the Perseverance rover to Mars. It is the first of its kind in terms of rotor craft design that is made for technological demonstrations. A successful flight of the Ingenuity helicopter will prove the feasibility of a powered flight in the thin atmosphere of Mars. The research and development work done for the Ingenuity helicopter has not only enabled detailed analysis studies in the field of low Reynolds number aerodynamics but also focuses on the overall vehicle design. The work done in the field of UAV flight for Mars has made a solid base for the future vehicle designs in terms of thin atmospheric flight as well as the autonomy of the vehicles. The Ingenuity helicopter is a kind of UAV that flies on the concept of helicopter mechanics. It has two rotors that are mounted co-axially on top of each other. Both rotors spins in opposite directions and this allows the vehicle to balance the torque produced by the rotors. Having co-axial rotors on the Ingenuity helicopter improve the way a small rotor can produce lift.

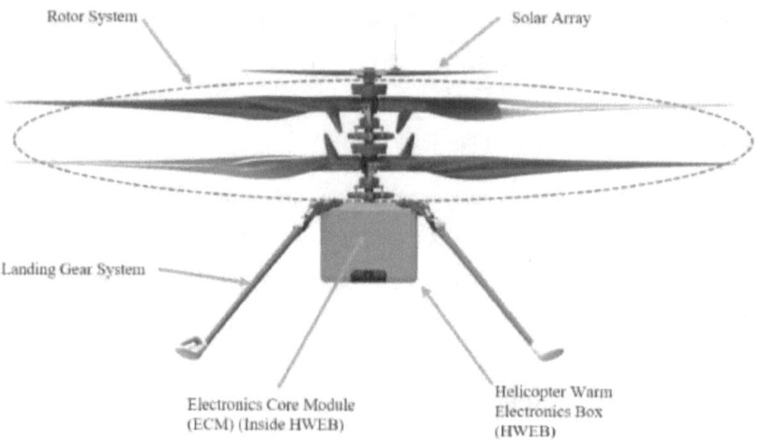

Figure 3.6: Ingenuity Helicopter

Because of more than one rotors it approximately doubles the lift produced by the rotor for the same amount of rotor disk area and it also helps the turbulent region of the flow to shift downwards. The schematic of the Ingenuity helicopter is shown in Figure 3.6 (Courtesy of Aerovironment and JPL [1]). The Ingenuity helicopter is powered by solar arrays that are mounted on the top of the rotor system. Because the overall mass of the system is only around 1.8 Kg, the required size of solar array in order to power the helicopter is small and therefore the solar array on top of the rotor system. The blades of the Ingenuity helicopter are specifically designed for the thin atmosphere of Mars. In Figure 3.6 the varying angle of attack for the rotor blades can be seen. The blade profile and its twist distribution is designed in such a way that the free stream velocity increases towards the tip of the rotor blades but the tip mach number still stays subsonic. The flat geometry of the rotor blades can be understood by the role it has to play in order to minimize the free stream velocity and restricting the tip speeds in the subsonic region [1].

3.1.4 Performance Evaluation of Ingenuity Blade Profile

The two dimensional analysis done for the considered airfoils for the Ingenuity helicopter reflect that the Reynolds number for the average displacement thickness takes place at the location where the laminar separation occurs. Therefore it can be concluded that the laminar to turbulent transition does not occur in the boundary layer reattachment phase. [19],[20]. These observations are also coherent with the findings of Lissaman [21], Mueller [22] and Carmichael [23]. In addition to this, the airfoils that were considered for the Ingenuity helicopter project, reflect the behaviour of laminar separation is delayed and happens in the downstream region of the flow. This is observed at around 80% of the chord length of the airfoil section. This further decreases the possibility of reattachment and turbulent transition of the flow on the upper surface of the airfoil.[6]

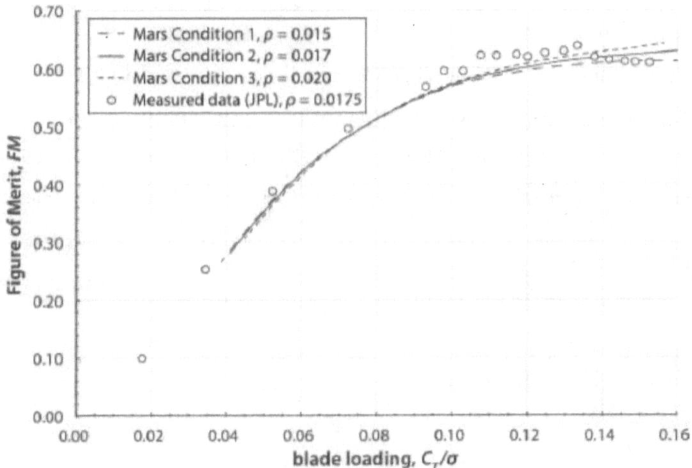

Figure 3.7: Performance of Ingenuity helicopter test model for different atmospheric conditions

There have been number of test articles fabricated for the testing of the blades of the Ingenuity helicopter. The JPL space simulator test facility is generally used to replicate space or planetary environments on earth in order to test the models for its future operations. The Ingenuity helicopter test models were tested at different Mars atmosphere conditions (using slight variation of atmospheric density ρ).

The tests that were carried out in the space simulator or vacuum camber were done at specific Mars conditions and those atmospheric conditions values are mentioned in the Table 3.2 (courtesy of JPL and Konning [6]. The flight performance of the Ingenuity helicopter was plotted against

Parameter	MARS	EARTH
Density, ρ (Kg/m^3)	0.017	1.225
Static Pressure p (Pa)	720	101325
Temperature T (K)	223	288.20
Gas constant R (m^2/s^2/K)	188.90	287.10
Dynamic viscosity μ (Ns/m^2)	$1.130 \cdot 10^{-5}$	$1.175 \cdot 10^{-5}$
Gamma γ	1.289	1.4

Table 3.2: Mars conditions for test articles

the blade loading of the vehicle. These tests were done at the previously mentioned atmospheric conditions but each time with slight variation in the atmospheric density value. The purpose was to observe the behaviour of the vehicle at slight variation in density to observe the density dependence of the vehicle. The results of vehicle performance with respect to the blade loading can be seen in Figure 3.7 (courtesy of Konning [6] and Konning [19]). The data suggests that the figure of merit of the Ingenuity helicopter vehicle fits well with the experimental data and the simulation data.

Chapter 4

Two Dimensional Flow Analysis

4.1 Experimental Setup

This section of the book will discuss about the 2D analysis carried out by [19] and [24] and will compare the simulation results from overflow to the analysis done by [18] and [16]. In this section, the considered boundary conditions and flow parameters will be discussed for the Ingenuity helicopter rotor model. The 2D airfoils were analyzed with the help of structured grid lines and solved with the help of Reynolds Averaged Navier Stokes equations in the OVERFLOW solver [25]. The performance of the airfoils are analyzed using the Martian conditions mentioned in Table 3.2 and the averaged Martian atmosphere conditions. The Prandtl number relates the momentum transport of the flow with its thermal transport property. The value of Prandtl number for the turbulent flow is considered as same as the Prandtl number for Air. The free stream turbulence is believed to have minor influence on the performance of the flat plates compared to the conventional airfoils. For the experimental setup the intensity value of the free stream turbulence is is assumed as 0.082% [6]. Even though in the 3D analysis the flow is considered as incompressible, in the 2D experimental setup it is assumed that the effect of the compressibility behaviour of the flow is very small.

The expected mach number range and the angle of attack of each blade section are mentioned in the Table 4.1 (courtesy of Konning [6]). One expected behaviour is noted here that the expected mach number range of 0.20 to 0.90 occurs at the last radial station of the set up that means the possibility of transitioning from subsonic to supersonic flow is more towards the tip of the rotor compared to the hub of the rotor.

CFD station	r/R	α[degree]	Mach M	$Re/M[10^{-4}]$
Station 1	0.091	-15 to 20	0.10 to 0.30	1.074
Station 2	0.200	-15 to 20	0.10 to 0.40	2.984
Station 3	0.295	-15 to 20	0.10 to 0.50	4.176
Station 4	0.390	-15 to 20	0.10 to 0.50	4.176
Station 5	0.527	-15 to 20	0.20 to 0.50	3.451
Station 6	0.762	-15 to 20	0.20 to 0.70	2.564
Station 7	0.924	-15 to 20	0.20 to 0.85	1.825
Station 8	0.991	-15 to 20	0.20 to 0.90	0.724

Table 4.1: Angle of attack (AOA) and Mach numbers for different radial stations

4.1.1 Turbulence and Laminar Separation Bubble Expectations

The modelling of turbulent flow at low reynolds number shows irregular conclusions drawn by different theories. In a study, Kunz [26] expected the flow to be almost completely laminar in the reynolds number range of 10^4 for Micro Aerial Vehicle performance calculations. Schmitz expected that the reynolds number range of 10^4 is good enough to change the shear layer to the turbulent flow when considered for a flat plate analysis [7]. In a different study, laminar reattachment and laminar separation were observed at the angle of attack of $3°$and the precise reynolds number of $1.0 \cdot 10^4$. And when the reynolds number was increased to the value of $5.0 \cdot 10^4$, the transition in shear layer was observed which led to the turbulent reattachment [27],[11].

The complexities of the laminar separation bubble might only be properly understood using Direct Numerical Simulations. On the other hand, multiple transition and turbulent models which use Reynolds averaged navier stokes and large Eddy Simulation [28],[29] have showcased promising results for the considered range of reynolds number. The approach of Large Eddy Simulations have shown promising results for the range of low reynolds numbers and high mach number as compared to the approach that uses reynolds averaged navier stokes equations. This is shown well in work done by Anyoji [18] for the performance evaluation of thin plates.

4.1.2   Definition of 2D Geometry

For the setup the flat plate and a cambered airfoil were considered to be composed of a mesh grid. This allowed the solver to optimize working area of the flow simulation and to focus the simulation effort towards the geometry defined inside the geometry.

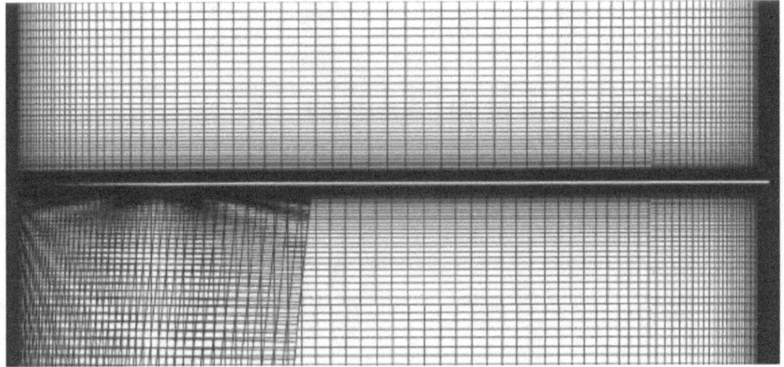

Figure 4.1: Geometry of flat plate defined in grid

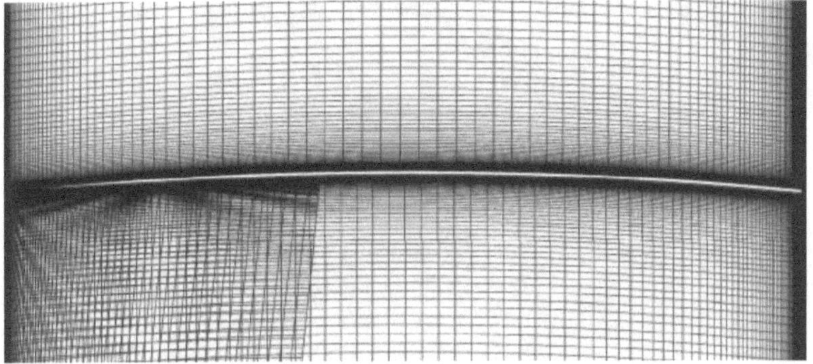

Figure 4.2: Geometry of cambered plate defined in grid

The grid was modified in order to define the beveled leading edge parameter (sharpness). The geometries of the flat plate and cambered plate can be seen in Figure 4.1 and Figure 4.2 (courtesy of Koning [6]). The far field limit for the geometric definition is set to be 50 times the chord length of the flat or cambered plate. In aerodynamics, it is a basic understanding that the lift characteristic is improved for a cambered airfoil compared to a symmetrical airfoil. For this reason the thickness considered is only about 1% of the chord length [30],[31]. The shape of the trailing edge is not smoothed because it is assumed that the trailing edge does not have much influence on the 2D analysis performance. From the past experiments [30] it is suggested that the camber value for the cambered plate is most appropriate is kept around 4 to 6%.

4.2   Result Validation of 2D Analysis

The simulations carried out by Koning[6] were compared against the predictions of studies by [16] and [32]. The significant part of the results are validated against the results from Laitone [16] and Okamoto et al. [32]. As mentioned in the previous section, the analysis was carried out using the OVERFLOW solver and part of the results are also validated against the data from the solver.

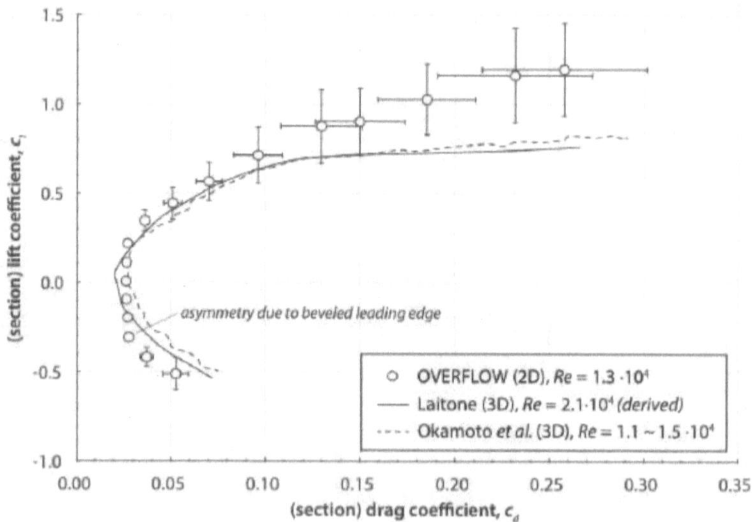

Figure 4.3: Cl versus Cd for a flat plate

The results from OVERFLOW solver are compared against Okamoto [32] and Laitone [16] for a flat plate for the low reynolds number. This is shown in Figure 4.3 (courtesy of Koning [6]). For the analysis, Okamoto et al. used a plate that has thickness to chord ratio of 1% and Laitone used a plate with thickness to chord ratio of 4% for trailing edge and 1% for the leading edge. The lift and drag coefficients are also plotted against the angle of attack and are displayed in Figure 4.4 (courtesy of Koning [6]).

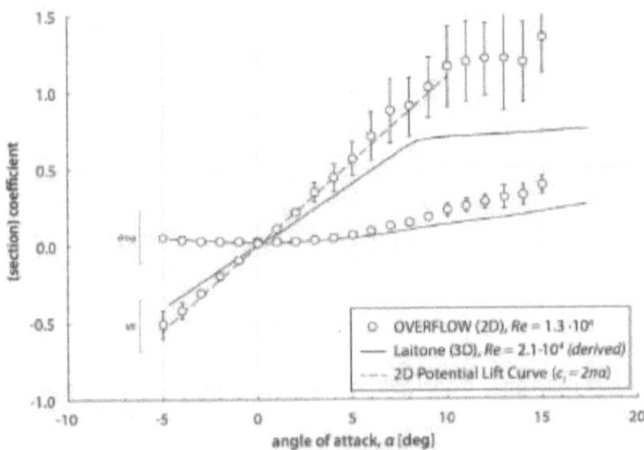

Figure 4.4: Angle of attack versus Cl and Cd for flat plate

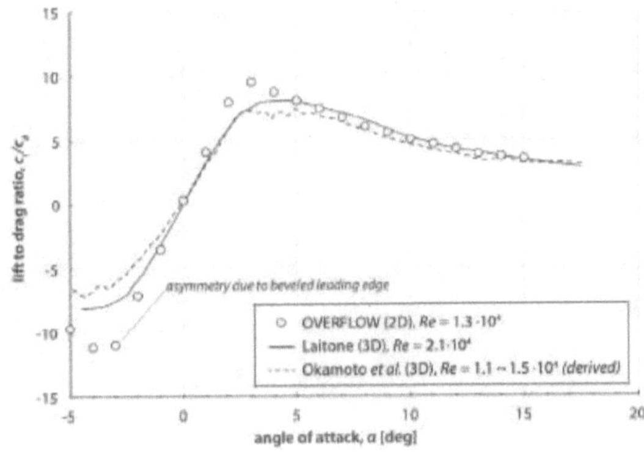

Figure 4.5: Lift/Drag coefficients versus Angle of attack for flat plate

The contradiction in performance is understood from the Figure 4.5 (courtesy of Koning [6]) because the performance at positive angles of attack is increased as compared to the negative angle of attack specially around 3.5°. This results were obtained for 2D analysis for flat plate from OVERFLOW where as results from Laitone were from 3D analysis for a flat plate.

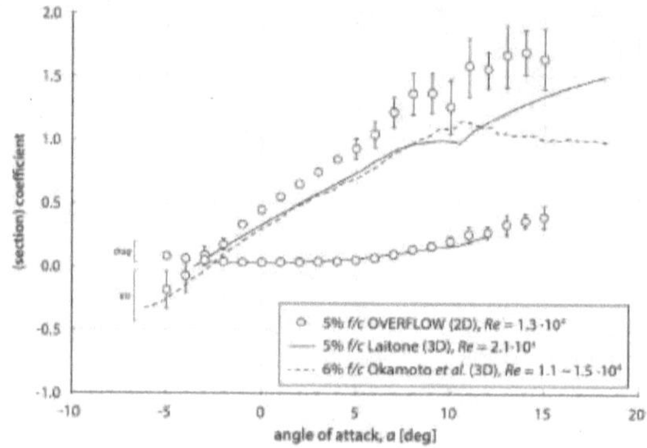

Figure 4.6: Angle of attack versus Cl and Cd for cambered plate

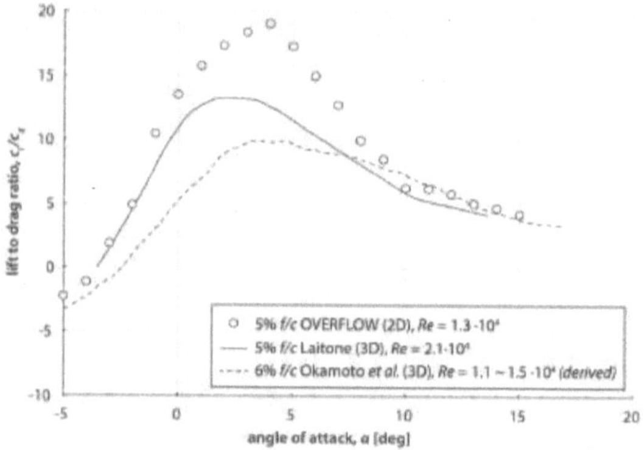

Figure 4.7: Lift/Drag coefficients versus Angle of attack for cambered plate

The analysis done by the OVERFLOW solver is plotted against the results from Laitone and Okamoto for a cambered plate in the Figure 4.6 and Figure 4.7 (courtesy of Koning [6]). The data from drag coefficients fit very well with the data from OVERFLOW and are not contradictory like in the case of flat plates. From the Figure 4.6 and Figure 4.7 it is clear that the changes in lift and drag coefficient become visible at angle of attack around 5°for the case of cambered plate.

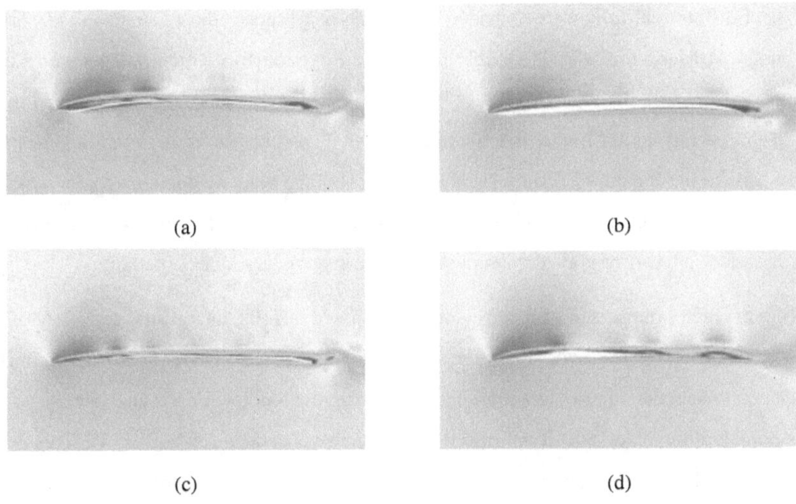

Figure 4.8: (a) Velocity profile over cambered plate airfoil (b) Velocity profile over ACA airfoil (c) Velocity profile over a DEP airfoil (d) Velocity profile over PAT airfoil

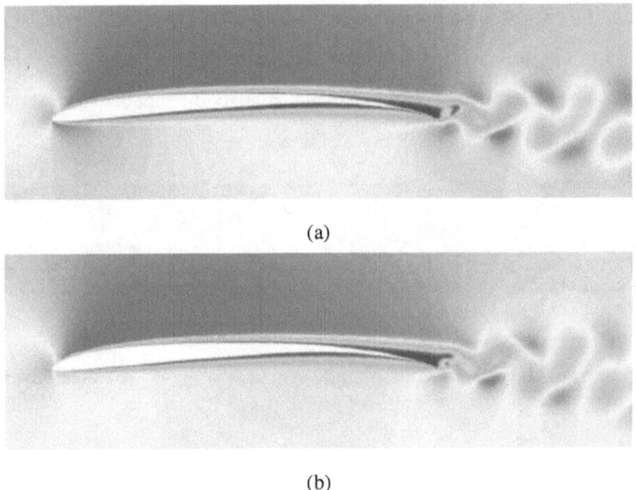

Figure 4.9: (a) Laminar boundary layer at angle of attack $\alpha = 0°$ (b) shear layer separation Angle of attack $\alpha = 2°$

Number of different airfoils were considered for analysis purpose. the ACA stands for Arbitrary Continuous Airfoils. the DEP is Double Edged Plate type airfoil which has geometry derived from flat plate a irfoil. the PAT airfoils are Polygonal A irfoils that has geometry derived from the wedged shaped airfoils for higher mach number flows. The coherent vortex shedding is observed in all airfoils in the Figure 4.8 (courtesy of Koning[24]). The velocity contours for the considered camber plate (CP), ACA, DEP and PAT airfoils have been plotted in Figure 4.8. Each plot has a slight variation in terms of vortex shedding and laminar boundary layer separation.

From the geometry derivation of the previously discussed airfoils, an optimal geometry of airfoil profile w as c hosen f or t he t ests f or I ngenuity h elicopter p roject. S ince t he a irfoil optimization were done in tools that are not available for public use, the studies for 2D analysis are compared in this book and the blade airfoil optimized for Ingenuity is chosen for Mars UAV. This is mainly because the analysis for airfoil are already carried out by JPL and this book is motivated from the Ingenuity concept. The separate analysis done for the ingenuity helicopter blade tip airfoil include the velocity contours, attachment of laminar boundary layer and shear layer separation behaviour. There have been other studied carried out but for the research aim in the CFD part of this book the study review only focuses on the mentioned states of analysis. The Analysis for the laminar boundary layer attachment was carried out for number of different angles of attack, and in order to see the variation because of small angle of attack change, the plot has been shown in Figure 4.9 (courtesy of Koning [24]) for a 0 and 2°angle of attack.

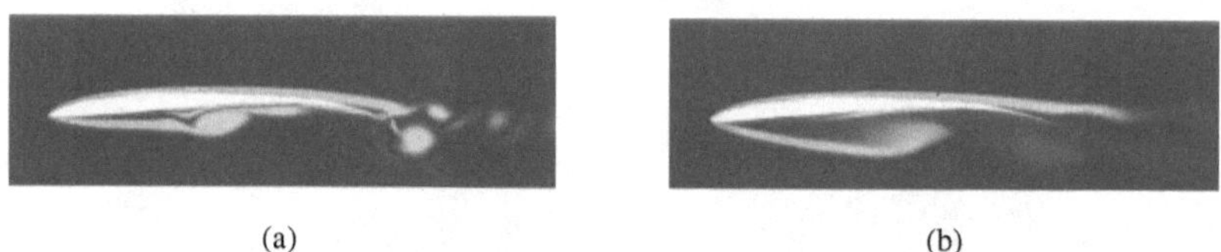

(a)   (b)

Figure 4.10: (a) Shear layer behaviour at angle of attack $\alpha = -2°$ (b) shear layer behaviour at Angle of attack $\alpha = -3°$

The movable pitch design of the rotor crafts needs analysis of the blade profile at negative angles of attack as well. At the hover conditions, the rotor blades were set at zero angle of attack. For the case of rotor blade pitch change, one of the rotor blade would be at negative angle of attack and the other will be at positive angle of attack. In order to see the flow behaviour for the negative angle of attack, the velocity contours were plotted as shown in Figure 4.10 (courtesy of Koning[24]). The separation of shear layer from the stream lines have been observed in Figure 4.10 (a) and as the blade angle is increased on the negative scale the separation becomes more prominent but the vortex shedding is reduced.

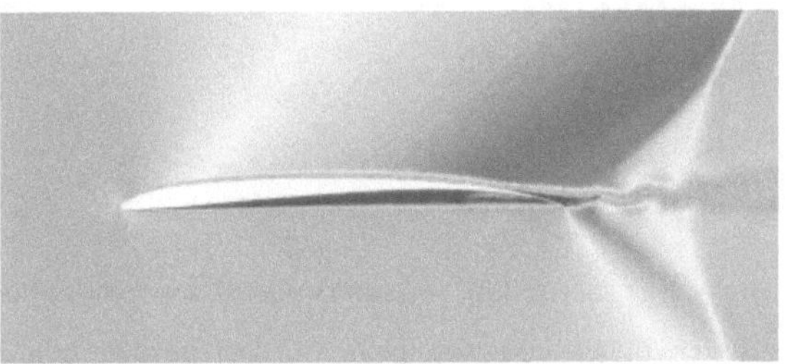

Figure 4.11: Extreme mach number in subsonic range showing shock wave formation

Because of the proportional relation between mach number and radial distance of station (the distance from the root to tip along the blade), the mach number increases as the distance of the considered airfoil section in increased from the rotor hub. Up until 75 to 80% of the rotor radius it is still well within subsonic limit but as the distance of CFD station increases beyond 80% of the rotor radius, the chances of flow becoming supersonic also increases. Therefore for the thin airfoils the analysis was necessary to be performed at higher mach number around 0.90. The Figure 4.11 (courtesy of Koning [24]) shows the formation of mach cone as the flow mach number is crossed beyond 0.90. This feature is more prominent towards the trailing edge and the effect of this is drastic decrease in rotor lift performance. The analysis was done for the extreme case with a considerable factor of safety. The angle of attack for the analysis beyond mach number 0.90 was takes as $3°$. The algorithms have shown an increase of 16% to 29% in lift performances for the mars helicopter technology demonstrator test articles [24],[33].

Chapter 5

Fluent Simulation 3D

5.1   Co-Axial Rotor System

A coaxial rotor system is characterized by two rotors placed one rotor above the other as shown in Figure 6.1. In the system of coaxial rotors, the flow of one propeller influences the lift creating ability of the other rotor. For a constrained area to put the rotor blades in order to generate additional thrust (compared to quadrotor), the co-axial setup is is efficient in terms of space occupied and the thrust produced. The Ingenuity helicopter [33] also uses a set of two rotors that are placed co-axially with the ability to do pitch control of each rotor separately. The model designed for this book project also has counter rotating co axial rotors, but instead of two rotors it uses eight rotors in total. The momentum theory of rotors also describes the performance optimization of co-axial rotors. Though the momentum theory is a 1D approach and is validated based on the power and thrust analysis for co-axial rotors. The approximations made in momentum theory calculates power loss that is introduced because of the two propellers and compares the performance against the thrust response. The comparative thrust of co axial system is comparatively higher because the motor speeds are constrained by the maximum value that will be analysed in this simulation.

The idea of using co-axial rotors in this book it not only aimed at increased thrust but also it compliments on the simplicity of the rotor system. The Ingenuity vehicle uses the concept of helicopter flight mechanics in order to manoeuvre where as the proposed vehicle for this book is a quadrotor with co-axial rotors and as such moves by controlling the speed of the four rotors. This means that the Mars UAV does not need the variable pitch control in order to perform roll, pitch and yaw. This simplifies mechanical system as well as control architecture to change pitch of the rotor blades.

5.2   Simulation Setup In ANSYS Fluent

The rotor system used for the Mars UAV is designed in Fusion 360 CAD designing software and is converted into mesh body to analyze in ANSYS. ANSYS is a software that contains different modules that are designed to carry out fluid simulations, structural loading, thermal simulations, mesh creation etc. Because part of the goal of this book is to do fluid simulation for rotor blades for a Mars UAV, the Fluent module was chosen to carry out the simulation task. Fluent uses an step file of a cad design of the part, model or system that needs to be analyzed. The rotor blade was imported from fusion and simplified in the design modeler of Fluent module. In order to do the fluid simulation, it is required to define the bounding box that contains the enclosure volume of the rotor system. The rotor system is enclosed in the static part of the bounding box and when the base parameters of atmospheric data is applied for the simulation, the software applies those parameters as data for static part of the volume.

There are two goals that needs to be addressed through this fluid s imulations. The first goal is to extract the thrust for the system of co axial rotors. The second aim is to observe the flow behaviour around the solid body of the rotor blades. The interaction of flow l ines w ith s olid b ody gives insights into the transition phase of the flow at different CFD stations (CFD stations are considered as locations from the hub of the rotor, out in the radial direction towards the tip of the rotor). The analysis is also done at number of different rotation speeds to observe the turbulence nature of the flow when the revolutions are i ncreased. The thrust is calculated as the pressure force applied by the fluid on to the rotor solid body s urfaces. The total thrust that w ill be displayed in the plots is the sum of the thrust force on upper rotor and the lower rotor. Once the bounding box is created in the design modeler, the rotor design is imported into the mesh module. In this section different mesh were created to optimize the simulation outcome. The best mesh quality is fine and the least precise used is the coarse mesh. Because of the limited computing power of the machine, the very fine mesh was not attempted.

5.3   Simulation Results and Validation

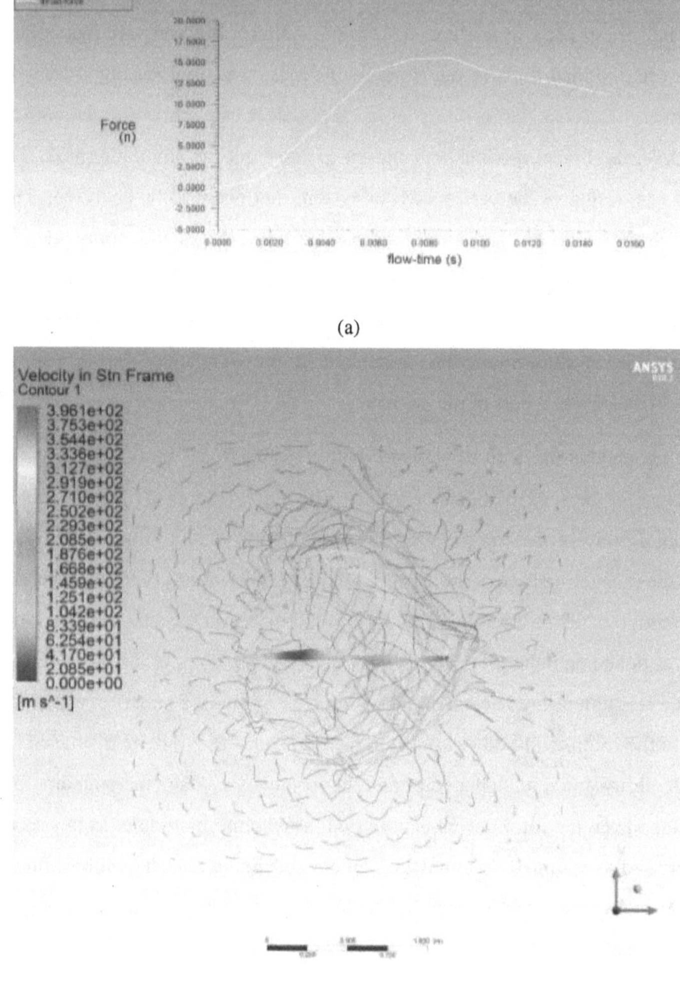

(a)

(b)

Figure 5.1: (a) Thrust plot for a single rotor (b) Velocity contour with flow lines for a single rotor

Once the mesh is created around the solid body, the body is converted into the form which can be interpreted by the solver in order to define the contact regions on the surface. In order to compare the flow behaviour between single and coaxial rotors, the first attempt was made to do the simulation only for a single rotor. In this case of the simulation, the rotor spins at 3200 rpm. In Figure 5.1 (a) the thrust force is plotted against the flow run time. After the flow convergence, the thrust value at 3200 rpm is observed to be around 11.5 Newtons. The thrust value is high considering the low atmospheric density and surface pressure used for the simulation (Mars conditions). On the other hand in order to reach this thrust value we have provided high rpm value to the rotor. And the effect of the high rpm for a rotor of this size is observed in Figure 5.1 (b). The turbulent nature of the flow and disturbances in the stream lines is easily seen towards the tip of the rotor. The speed of sound on Mars is given by equation (5.1).

$$a = \sqrt{\gamma R T} \tag{5.1}$$

For average surface temperature on Mars, the speed of sound is computed to be around 244 m/s. In the Figure 5.1 (b) the velocity in stationary frame is observed to be around 396 m/s at the tip of the rotor. From equation (2.19) the mach number for these values in Mars condition is calculated and the mach number is 1.62 and this is clearly not subsonic flow. Thus the turbulent behaviour of the flow around the tip is the influence from the shock wave that is generated at the tip of the rotor blade. From this it can be concluded that the high value of thrust can be achieved by increasing the rpm but at the cost of having a supersonic flow. Also once the flow becomes supersonic, the thrust decreases exponentially until the RPM is not lowered. And this limits the operation of the Mars UAV. In order to understand the importance of using co axial rotor for the Mars condition, the simulation of one rotor for a higher rpm was carried out.

For the co-axial rotor simulation, both rotors are at zero angle of rotation at the beginning of the simulation. The first set of simulations were carried out at 3000 rpm. The CFD data extracted from the solver setup was post processed in the CFD post processor of Fluent and the different contours of velocity, pressure and turbulence kinetic energy was plotted in graphically readable manner.

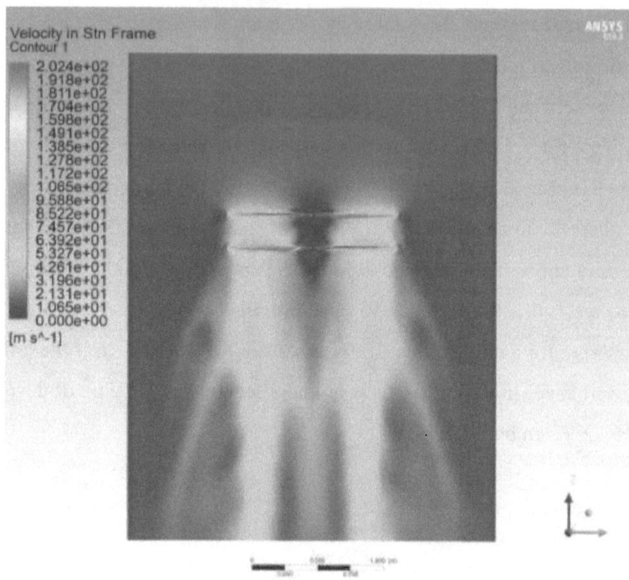

Figure 5.2: Velocity in stationary frame contour for co axial rotors

The Figure 5.2 is the contour of the velocity measured in static as well as rotating domain of the simulation. The use of two rotors mounted co-axially allows to decrease the revolutions per minute considered for the rotors while at the same time increasing the overall thrust of the system. In Figure 5.2 it can be observed that the maximum velocity value at any point inside the considered bounding box is not going beyond 202.4 m/s. This means that even at the tips of the rotor blade, the velocity is limited inside the 202.4 m/s value. This implies that the mach number in this case is less than 0.82. Therefore the flow is always in the subsonic region and the shock waves do not form in this case. The wake is generated in this case because of the influence of the upper rotor on the flow around the lower rotor but the effect of wake is not prominent in this case because of the smooth geometry of the rotor and high performance cambered airfoil used to model the blade.

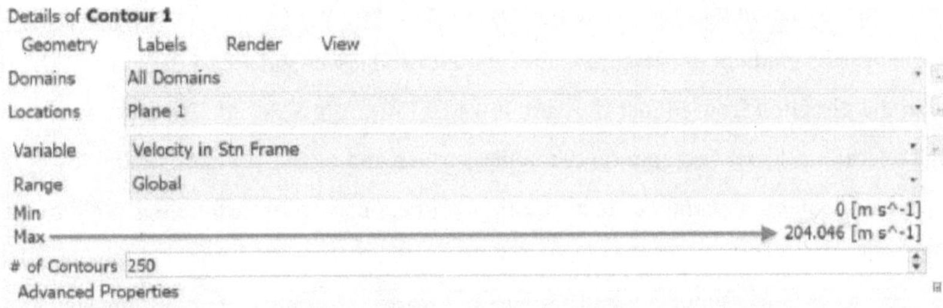

Figure 5.3: Maximum velocity value in stationary frame for 3000 rpm

The validation of the maximum velocity for the co-axial rotor system case is shown in Figure 5.3. This validates the idea of using co-axial rotors in the rpm range of less than 3000 rpm for a performance that does not include shock waves and supersonic flow turbulence.

```
Forces - Direction Vector (0 0 1)
                         Forces (n)
Coefficients
Zone                        Pressure        Viscous        Total
Pressure         Viscous        Total
propellers-lower_rot        7.4729266      -0.022771221    7.4501554
747.29268       -2.2771221        745.01556
propellers-upper_rot        8.245266       -0.020343466    8.2249225
824.52661       -2.0343466        822.49227
----------------------------  ------------------  --------------------
----------------------------  ------------------  --------------------  --------------------
Net                         15.718193      -0.043114686    15.675078
1571.8193       -4.3114687        1567.5078
```

Figure 5.4: Force report from Fluent

The Figure 5.4 shows the force report from Fluent. This includes Pressure and viscous force. The dominant force is the pressure force which is responsible for the thrust. The force is calculated on upper and lower rotor separately based on the Mars conditions and rotor blade geometry. The total force is around 15.67 Newtons and that is taken as thrust force for the vehicle design optimization purpose.

The basic idea of lift generation by the airfoil and eventually by a rotor blade is based on the pressure distribution and pressure difference. It is important to discuss the pressure contours in order to get better understanding of the lifting performance of the rotor system. The airfoil generates lift on the principle of gradient in pressure value on upper and lower surface of the airfoil. The force is created in the direction from higher pressure towards lower pressure side. In order to generate lift force the airfoil has lower pressure on the upper surface and higher pressure on the lower surface. In the pressure contours of the co axial rotors this phenomenon of differential pressure is easily identified. The Figure 5.5 shows the pressure field inside the bounding box volume considered for the analysis. As discussed above the difference in pressure is quite evident in Figure 5.5. The red part on the lower side of the rotor indicates higher pressure and green part in the contour on the lower side of the rotor indicates lower pressure compared to the upper surface. This phenomenon is same for both upper and lower rotor despite both rotors spinning in opposite directions.

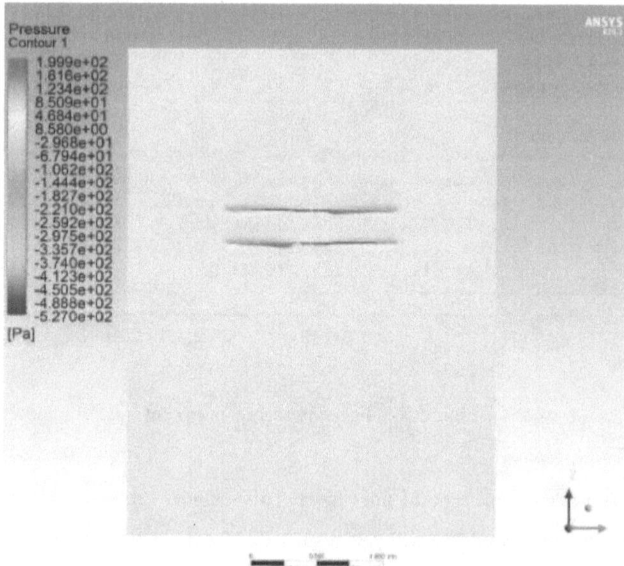

Figure 5.5: Pressure contours for co axial rotors

In order to see the effect of of turbulence inside the control volume, the flow is also also plotted in terms of streamline and their velocity. This helps to visually see the effects of turbulence and corresponding velocity in global frame. Figure 5.6 shows the streamlines flowing in from positive Z direction. In the contour, the velocity is never reaching beyond 204 m/s mark. This also supports the evidence about absence of shock waves towards the tip of the rotor. If Figure 5.1 (b) is compared with Figure 5.5, the observed differences are more stream line flow and less turbulent behaviour in the case of co-axial rotors. As discussed previously, the wake is generated but the effect of wake is not seen to affect the flow characteristics.

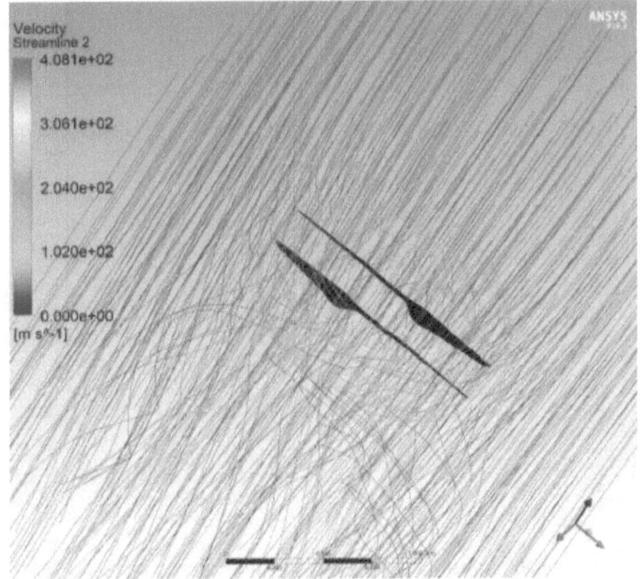

Figure 5.6: Global velocity contour with streamlines

Since the Figure 5.6 display the stream line velocity in global frame, the surface of the propellers are not colour coded here. This means that for this part of post processing the surfaces of the propeller blade were considered as not sensitive to pressure. This was done to enhance the understanding on flow behaviour around the rotor and inside the rotating module of control volume.

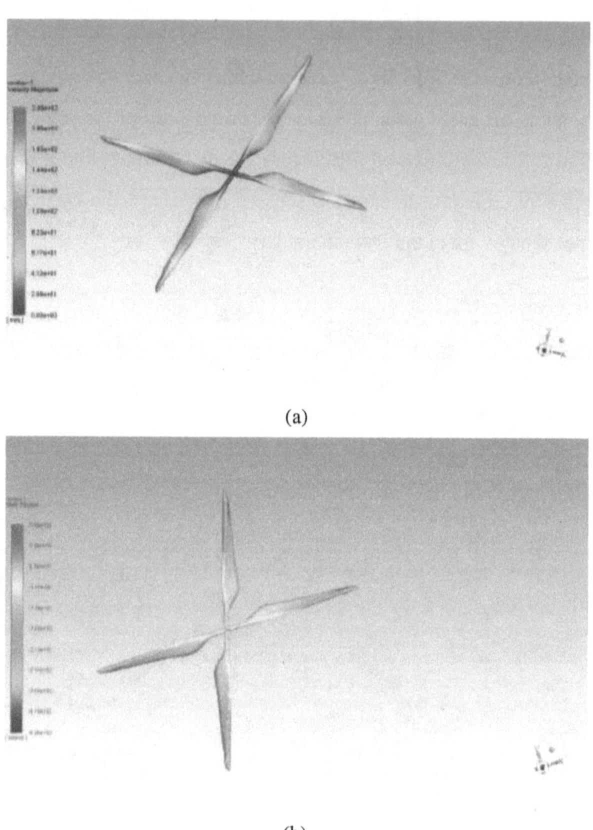

(a)

(b)

Figure 5.7: (a) Velocity contour lines on rotor surfaces (b) Pressure contour lines on rotor surfaces

The negative pressure on the upper surface of the both rotors can be seen in Figure 5.7 (b). Although the pressure values showed in Figure 5.7 (b) are quite low, the idea of having negative pressure on the upper surface drives the thrust force concept. The values are very low because the reference values considered for the simulation are the atmospheric parameters of Mars. As a working fluid CO_2 is considered with only about 1% of that of Earth.

Chapter 6

CAD design of Mars UAV

In this chapter the 3D design of the Mars UAV model is discussed. The Mars UAV is a vehicle that is inspired from the conventional quadrotor system but it is modified in order to meet the thrust requirement in the thin atmosphere of Mars. The Mars UAV system is designed with the purpose of making the model that is able to withstand Martian conditions like dust storm and temperature changes between day and night.

The isometric view of the Mars UAV CAD model is shown in Figure 6.1. The model is designed in Autodesk Fusion 360 software. The software has capabilities to design, render, animate and simulate the model in the same environment. Because of the detailed simulation required for this book project, the flow simulation is not carried out in the same software because of the issue of computational power. The model was exported as an step file to the Fluent in order to simplify the model and simulate for Mars conditions.

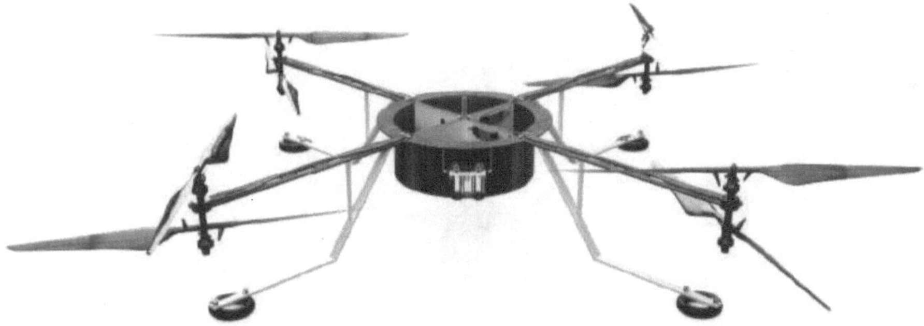

Figure 6.1: Isometric view of the Mars UAV model

In order optimize the dimensional parameters of a UAV mode while making the CAD design, it is necessary to have base reference values of some part on which the dependent model dimensions can be derived. In the case of Mars UAV, the rotor blade size is taken as the base reference. The fluent simulation was done with only the rotor blades so by rough estimation of the thrust values required, the diameter dimension of the rotor blade can be optimized. In the case of Mars UAV, as discussed in Figure 5.4, in order to generate around 15 Newtons of thrust with the set of co axial rotors, the rotor blade diameter should be around 1125 mm. The rotor diameter value might seem too large for general consideration but the thin atmosphere of Mars and low efficiency of lift producing capability, the large value of the rotor diameter is justified. The design of the Mars UAV rotor blade is shown in Figure 6.2.

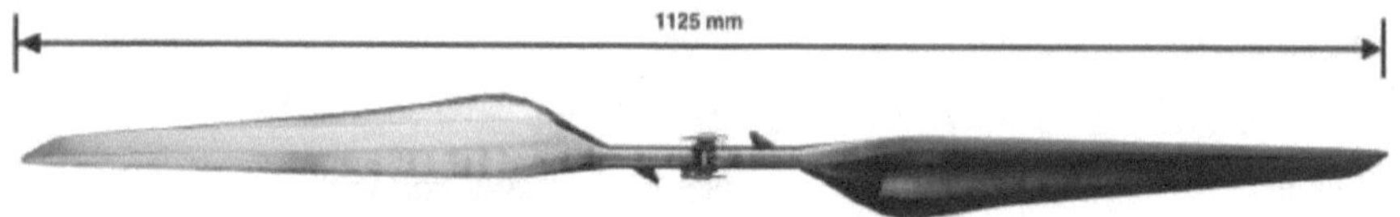

Figure 6.2: Rotor blade of Mars UAV

The isometric view of the rotor blade is shown in Figure 6.3. The rotor blade is designed with varying pitch.

Figure 6.3: Side view : Rotor blade of Mars UAV

The Mars UAV model is equipped with landing gear which have small rubber tube kind of cushion at the bottom of it. Because rubber is flexible and can compress under load, this small part added to the landing legs helps in absorbing significant load in case of hard landing. Also the shock absorber is made of rubber tube so it does not need any additional system or power to work as shock absorber. It is an independent addition to the landing system which is fastened to the legs of the system. The landing legs are designed in such a way so that the stiff structure of the landing legs is not coming in the way of the rotor disk. Despite of having large rotor diameter and a co-axial system with two rotors on each axis, the landing legs have enough space to assist in landing while not disturbing the flow from the rotors.

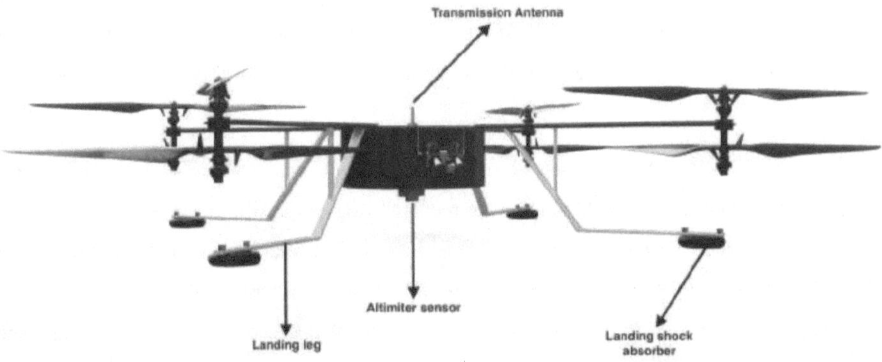

Figure 6.4: Subsystems Mars UAV model

The model of the proposed Mars UAV is equipped with a helical transmission antenna which transmits the signal to the nearby rover. This is inline with the mission objective of pre mapping the terrain for rover. The helical antennas are small in size but are also efficient in terms of consuming less transmission power. On the bottom surface of the payload bay, an altimeter sensor is mounted. This sensor measures the altitude reading and feeds it to the flight controller of Mars UAV in real time. An Inertial measurement Unit (IMU) is also included in the subsystem of the UAV in order to measure accleration and gyroscopic effects.

The Payload bay of the Mars UAV is divided into two sections. The upper section of the payload bay contains flight controller, Sensors, science instruments. While the lower section of the payload bay contains batteries. For the purpose of terrain mapping and path planning, a LiDAR sensor as well as a vision camera is mounted on the Mars UAV. The batteries are the instantaneous power source for the Mars UAV. Mars has very low temperature during the night time and in order to keep the batteries warm at night a separate heating system is required. This thermal management of the subsystems and batteries can be either active or passive. In active thermal management, the subsystems temperature will be maintained in the operating range via thin film h eaters. The thin film h eaters a re h eating d evices w idely u sed i n s pace i ndustry. These d evices c onsume minimal electrical power and provide a great range of temperature gradient. These heaters can be placed adjacent to the subsystem to conduct the heat. The passive way of thermal management includes CO_2 circulation. CO_2 is believed to be poor conductor of heat thus if CO_2 is circulated through channel inside the payload bay, it helps in trapping the heat that is produced by various subsystem within the payload bay. CO_2 is abundant in Mars atmosphere therefore this can be considered as an option.

The batteries of the system are designed to be powered from solar arrays. The design of the solar arrays is not included in the CAD model for reducing the complexity of the model for the scope of this book project. However, an extendable and retractable solar arrays can be mounted on the arms of the Mars UAV. These solar arrays are called roll out solar arrays. When retracted, the solar array can fold itself in a form of cylinder and when extended the solar array unfolds like a sheet of paper. This can be considered for the Mars UAV because the size of the arms of Mars UAV is large therefore an extendable solar array which is around 1300 mm × 1300 mm can be mounted. Moreover the UAV is a symmetric design so in total four extendable solar arrays can be mounted on the four arms of the Mars UAV. The large size of solar arrays is an advantage when we consider the solar to electrical power conversion efficiency on Mars is low. The solar arrays will be extended when the Mars UAV is resting on Martian ground in idle mode and these will be retracted when Mars UAV attempts to take a flight.

Chapter 7

Vehicle Dynamics

7.1 Conventional Quadrotor

A quadrotor is a kind of Unmanned Aerial Vehicle that is powered by four propulsion units. The four propulsion units are mounted on the four corners of the vehicle frame and the frame is structured in such a way that the payload can be placed close to the center of gravity of the vehicle. Generally the quadrotors can be controlled by a joystick controller in a manual way. The autonomy of the operation is necessary when the quadrotor system is assumed to fly out of sight of the operator or to make complex maneuvers for which the control response from manual operation is not fast enough. The idea of developing the Mars UAV is inspired from the advantages quadrotor systems have over helicopter vehicles. The helicopter need a complex mechanism in order to correct the pitch of the rotors when doing maneuvers. Whereas, quadrotors change their orientation just by changing the rotor speeds. All three, roll, pitch and yaw movements can be achieved just by giving specific commands to the motors in order to change rotor speeds and this does not involve any kind of mechanisms or mechanical control. However the downside of the Mars UAV system is that for large rotors the actuation effort is very large to speed up or slow down the motors and this also results in slow response time. Therefore for very large rotors, variable pitch is used since motors can not speed up/down quickly.

7.2 Quadrotor Dynamics

In order to define the equations of motion of a quadrotor, it is important to define the notations and frame of reference first. For a quadrotor, two frames of reference are considered. A world frame of reference which is always stationary and the other is quadrotor body frame of reference. The structure and motion of the vehicle are described in body frame of reference where as the position commands are given in world frame of reference. However these notations are only considered for defining the mathematical model and simulation purpose. In real life it is very difficult to define positions and reference in world frame in an unknown environment.

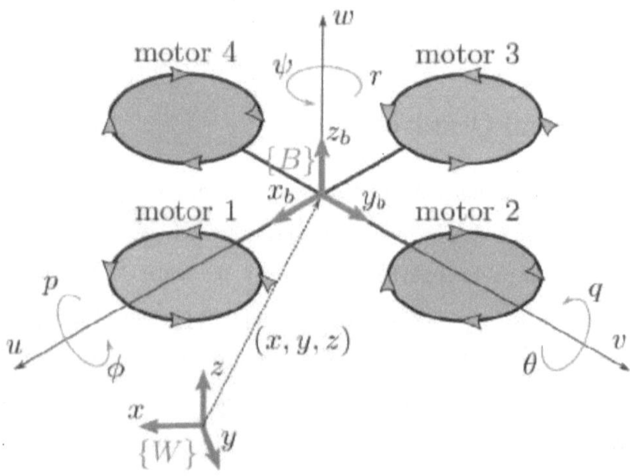

Figure 7.1: World and Body frame of reference for a quadrotor

The two frames of reference are as described in Figure 7.1 [34]. The control of the quadrotor is achieved by adjusting the motor speeds. Due to the difference in motor speeds, moments are generated because of the coupled forces from opposite motors. There are two configurations in which a quadrotor motion is defined. The configurations are $\times$ and $+$ configurations. In the definition of dynamic models of quadrotor and Mars UAV in this book project the $+$ configuration is considered. As shown in Figure 7.1 the opposite rotors spin in the same direction and the adjacent rotors spin in opposite direction in the quadrotor model. The six degrees of freedom motion of the quadrotor is defined in terms of forward and backward directions, lateral motion, vertical movement, roll, pitch and yaw movements. These motions can be narrowed down to three force moments that needs to be addressed in the control aspect of a quadrotor. in $+$ configuration of the vehicle, the roll, pitch and yaw moments are achieved by specifically altering the rotor thrust as described in Figure 7.2 [35].

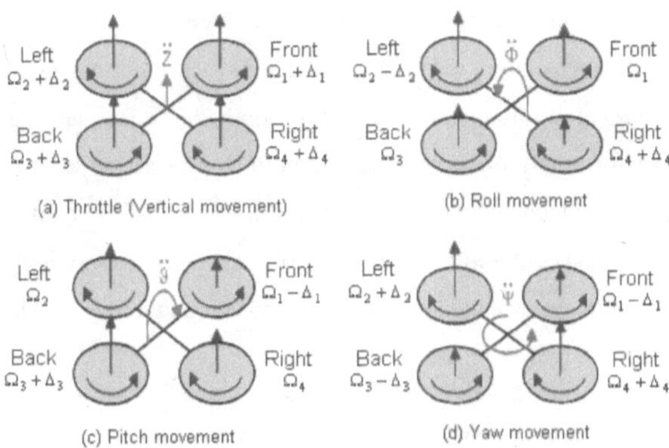

Figure 7.2: Differential thrusts for roll, pitch and yaw moments

When the four input motor speeds are same, the reactive torque is balanced from all four motors and the quadrotor stays in place. When the four input speeds are not exactly the same, the reactive torque from motors are different and depending upon the the difference in the speeds, the rotor will behave accordingly. When the speeds of all four rotors increase or decrease together with same amount, the quadrotor will either rise up or go down. Now there are four inputs as motor speeds and the outputs are six as it is 6 DOF system, the quadrotor is considered as underactuated system [36]. In order to make mathematical model of the quadrotor, some assumptions are made and those assumptions are:

1. Quadrotor is a rigid body

2. The quadrotor structure is symmetric

3. The ground effect is ignored

The quadrotor mathematical model is given by 12 output states. The x,y and z position coordinates, $\dot{x},\dot{y}$ and $\dot{z}$ are the translational velocities, ϕ,θ and ψ are the roll, pitch and yaw angles, $\dot{\phi}$, $\dot{\theta}$ and $\dot{\psi}$ are the angular velocities for roll, pitch and yaw moments.

7.2.1 Euler angles

Leonhard Euler introduced Euler angles that are used to define the orientation of a rigid body. In order to define an orientation in 3D euclidean space, the three parameters are needed. In this book project, the set of Euler angles used are ZYX Euler angles. The euler angles are also used in order to describe the orientation of reference frame relative to another reference frame and they are also implemented in transforming the coordinates of a point from one frame of reference to another frame of reference. The elemental rotation is defined as the rotation about the axis of frame of reference. The definition of euler angle represent three elemental r otations. The combination of rotation metricise used to define Euler angles are as shown in equations (7.1), (7.2) and (7.3) and in Figure 7.3 [36].

$$R_x(\phi) = \begin{bmatrix} 1 & 0 & 0 \\ 0 & \cos(\phi) & -\sin(\phi) \\ 0 & \sin(\phi) & \cos(\phi) \end{bmatrix} \tag{7.1}$$

$$R_y(\theta) = \begin{bmatrix} \cos(\theta) & 0 & \sin(\theta) \\ 0 & 1 & 0 \\ -\sin(\theta) & 0 & \cos(\theta) \end{bmatrix} \tag{7.2}$$

$$R_z(\psi) = \begin{bmatrix} \cos(\psi) & -\sin(\psi) & 0 \\ \sin(\psi) & \cos(\psi) & 0 \\ 0 & 0 & 1 \end{bmatrix} \tag{7.3}$$

The world frame coordinates and the body frame are co related by the rotation $R_{zyx}(\phi, \theta, \psi)$ and $c(\phi) = \cos(\phi)$, $s(\phi) = \sin(\phi)$, $c(\theta) = \cos(\theta)$, $s(\theta) = \sin(\theta)$, $c(\psi) = \cos(\psi)$ and $s(\psi) = \sin(\psi)$.

$$R_{zyx}(\phi, \theta, \psi) = R_z(\psi) \cdot R_y(\theta) \cdot R_x(\phi)$$

$$R_{zyx}(\phi, \theta, \psi) = \begin{bmatrix} c(\psi)c(\theta) & c(\psi)s(\theta)s(\phi) - s(\psi)c(\phi) & c(\psi)s(\theta)c(\phi) + s(\psi)s(\phi) \\ s(\psi)c(\theta) & s(\psi)s(\theta)s(\phi) + c(\psi)c(\phi) & s(\psi)s(\psi)c(\phi) - c(\psi)s(\phi) \\ -s(\theta) & c(\theta)s(\phi) & c(\theta)c(\phi) \end{bmatrix} \quad (7.4)$$

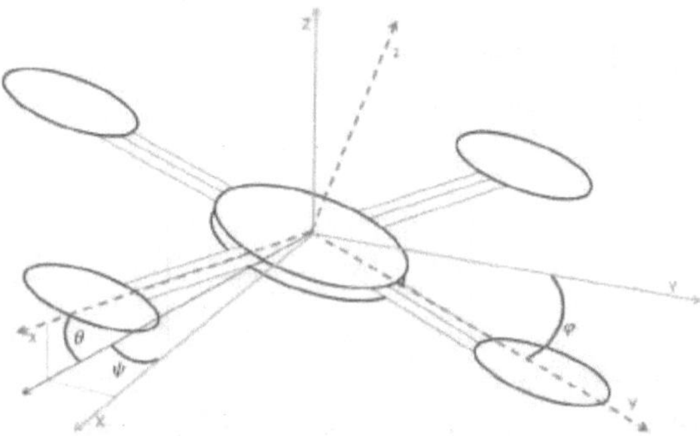

Figure 7.3: Euler angles representation

7.2.2 Mathematical Model of Quadrotor

In order to describe three dimensional motion of the quadrotor, a mathematical model of quadrotor is developed using the Newton and Euler equations. The aim of developing a mathematical model is to provide a dynamic model that is reliable in terms of simulating and controlling the quadrotor behaviour for 6 DOF motion. The linear and angular position vector is given as $[x\ y\ z\ \phi\ \theta\ \psi]^T$ and this position vector is with respect to the world reference frame. The linear and angular velocity vector is given as $[u\ v\ w\ p\ q\ r]^T$ and the velocity vector is with respect to the body frame of reference. For control purpose, in order to transform the velocities from body frame to the inertial frame of reference, the following transformation is considered. V_B is the velocity vector in body frame and V is the velocity vector in inertial frame of reference. Therefore,

$$V = R \cdot V_B \quad (7.5)$$

$V_B = [u\ v\ w]^T$ is transformed into the velocity in inertial frame and that is given by $V = [\dot{x}\ \dot{y}\ \dot{z}]^T$. In the same way the angular velocities can be transformed from the body frame to the inertial frame of reference.

$$\omega = T \cdot \omega_B \tag{7.6}$$

where $\omega = [\dot{\phi}\ \dot{\theta}\ \dot{\psi}]^T$ is the vector of angular velocity in inertial frame of reference. The Transformation vector T is defined as equation (7.7). [36].

$$T = \begin{bmatrix} 1 & s(\phi)t(\theta) & c(\phi)t(\theta) \\ 0 & c(\phi) & -s(\phi) \\ 0 & \frac{s(\phi)}{c(\theta)} & \frac{c(\phi)}{c(\theta)} \end{bmatrix} \tag{7.7}$$

From newton's second law of motion, The total force acting on the quadrotor body is represented as,

$$F_B = m(\omega_B \times V_B + \dot{V}_B)$$

Where $F_B = [F_x\ F_y\ F_z]^T$ is the force vector. Also the total torque applied to the quadrotor body is given as,

$$\tau_B = I \cdot \dot{\omega}_B + \omega_B \times (I \cdot \omega_B)$$

Where $\tau_B = [\tau_x\ \tau_y\ \tau_z]^T$ is the torque vector. **I** is represented as the moment of inertia matrix.

$$I = \begin{bmatrix} I_{xx} & 0 & 0 \\ 0 & I_{yy} & 0 \\ 0 & 0 & I_{zz} \end{bmatrix} \tag{7.8}$$

7.3 Mars UAV Model Dynamics

Mars UAV is a co axial rotor craft system that is inspired from the conventional quadrotor design. The goal is to produce more thrust by using a co-axial system of rotors. The quadrotor control is achieved by providing input to the four motor speeds whereas in the Mars UAV system, the control will be achieved by providing input to the eight motor speeds. The direction of rotation for each rotor for Mars UAV is as shown in Figure 7.4 [37].

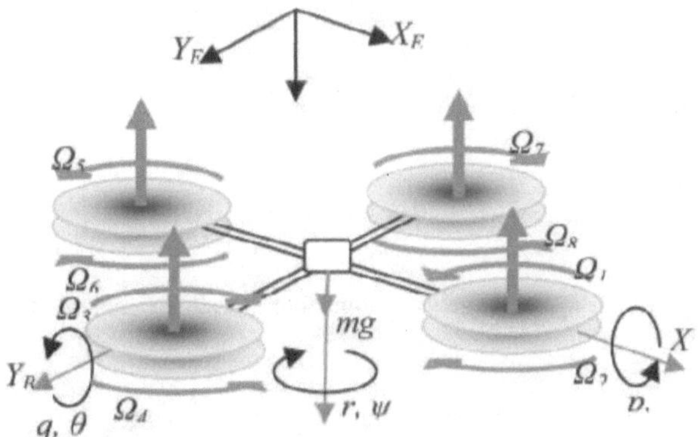

Figure 7.4: Mars UAV rotors representation

In order to balance the total torque of the vehicle, each pair of coaxial rotors spin in opposite direction with respect to the adjacent pair of rotors. Since the rotor is in + configuration, the forward direction is X axis on the body frame. The rotor speed is denoted as Ω and the total thrust in upward direction is denoted as T. Therefore,

$$T_i = K_T \Omega_i^2, \; i = 1, 2, ..7, 8$$

$g = 3.711 m/s^2$ is the gravitational acceleration on Mars, B is the aerodynamic friction coefficient and $T = \sum T_i$ is the total thrust in body frame. Considering the Newton's second law of motion in inertial coordinate frame,

$$m\dot{V} = \begin{pmatrix} 0 \\ 0 \\ mg \end{pmatrix} - R \begin{pmatrix} 0 \\ 0 \\ T \end{pmatrix} - BV \tag{7.9}$$

The distance from the rotor hub to the center of gravity of the vehicle is d therefore, The roll moment of the vehicle about X axis is given as,

$$U_2 = dK_T(\Omega_7^2 + \Omega_8^2 - \Omega_3^2 - \Omega_4^2) \tag{7.10}$$

The pitch moment of the vehicle about Y axis is given as,

$$U_3 = dK_T(\Omega_5^2 + \Omega_6^2 - \Omega_1^2 - \Omega_2^2) \tag{7.11}$$

If the coefficient of drag is K_D then, the yaw moment of the vehicle about Z axis is given as,

$$U_4 = K_D(\Omega_2^2 + \Omega_4^2 + \Omega_6^2 + \Omega_8^2 - \Omega_1^2 - \Omega_3^2 - \Omega_5^2 - \Omega_7^2) \tag{7.12}$$

The total thrust T is represented as U1 for throttle control in Matlab model. The overall rotor speed is represented as,

$$\Omega_r = -\Omega_1 + \Omega_2 - \Omega_3 + \Omega_4 - \Omega_5 + \Omega_6 - \Omega_7 + \Omega_8 \tag{7.13}$$

The total rotational accleration of the vehicle is described as,

$$I\dot{\omega} = -\omega \times I\omega + \tau \tag{7.14}$$

Here ω is the angular velocity vector in inertial frame of reference. If only thrust force is considered in inertial frame of reference then the total forces and torque matrix is given as,

$$
\begin{pmatrix} T \\ \tau \end{pmatrix} = \begin{pmatrix} -K_T & -K_T & -K_T & -K_T & -K_T & -K_T & -K_T & -K_T \\ 0 & 0 & -dK_T & -dK_T & 0 & 0 & dK_T & dK_T \\ -dK_T & -dK_T & 0 & 0 & dK_T & dK_T & 0 & 0 \\ -K_D & K_D & -K_D & K_D & -K_D & K_D & -K_D & K_D \end{pmatrix} \begin{pmatrix} \Omega_1^2 \\ \Omega_2^2 \\ \Omega_3^2 \\ \Omega_4^2 \\ \Omega_5^2 \\ \Omega_6^2 \\ \Omega_7^2 \\ \Omega_8^2 \end{pmatrix} = A \begin{pmatrix} \Omega_1^2 \\ \Omega_2^2 \\ \Omega_3^2 \\ \Omega_4^2 \\ \Omega_5^2 \\ \Omega_6^2 \\ \Omega_7^2 \\ \Omega_8^2 \end{pmatrix}
$$

Because Thrust coefficient K_T, Drag coefficient K_D and the distance from center of gravity, d are constant and positive values, The rotor speeds for input can be computed from the thrust required and torque required values. Therefore, the rotor speeds are formulated as shown in equation (7.15).

$$
\begin{pmatrix} \Omega_1^2 \\ \Omega_2^2 \\ \Omega_3^2 \\ \Omega_4^2 \\ \Omega_5^2 \\ \Omega_6^2 \\ \Omega_7^2 \\ \Omega_8^2 \end{pmatrix} = A^{-1} \begin{pmatrix} T \\ \tau_x \\ \tau_y \\ \tau_z \end{pmatrix} \tag{7.15}
$$

7.3.1 Equations of Motion

Considering that the assumptions made for quadrotor also hold for the Mars UAV, the equations of motion can be formulated using Newton-Euler concept. The origin of the body frame of reference coincides with the center of gravity of the vehicle. The linear accleration in x direction is given as,

$$
\ddot{x} = [\sin\psi\sin\phi + \sin\psi\sin\theta\cos\phi]\frac{U_1}{m} \tag{7.16}
$$

The linear accleration in y direction is given as,

$$\ddot{y} = [-\cos\psi\sin\phi + \sin\psi\sin\theta\cos\phi]\frac{U_1}{m} \tag{7.17}$$

and the linear accleration in Z direction is given as,

$$\ddot{z} = [\sin\theta\cos\phi]\frac{U_1}{m} - g \tag{7.18}$$

Similarly the rotational dynamics can be described by the angular accleration in roll, pitch and yaw direction. The angular acclerations are given as,

$$\ddot{\phi} = [\dot{\theta}\dot{\psi}(I_{yy} - I_{zz}) - J_r\dot{\theta}\Omega_r + U_2]\frac{1}{I_{xx}} \tag{7.19}$$

$$\ddot{\theta} = [\dot{\phi}\dot{\psi}(I_{zz} - I_{xx}) - J_r\dot{\phi}\Omega_r + U_3]\frac{1}{I_{yy}} \tag{7.20}$$

$$\ddot{\psi} = [\dot{\phi}\dot{\theta}(I_{xx} - I_{yy}) + U_4]\frac{1}{I_{zz}} \tag{7.21}$$

In equations (7.19) and (7.20), J_r is the total rotational inertia of the vehicle.

7.3.2 State Space Representation

The State Space representation is a mathematical model of the Mars UAV system as a set of input, output and state variables that are co related by first order differential equations.[38] In state space representation, state space is the space that consists of all state variables as its axis. The most general state space representation of a linear system is given as,

$$\dot{x}(t) = Ax(t) + Bu(t) \tag{7.22}$$

$$y(t) = Cx(t) + Du(t) \qquad (7.23)$$

In equations (7.22) and (7.23),

$x(t)$ is the 'State Vector'

$y(t)$ is the 'Output Vector',

$u(t)$ is the 'Input Vector',

A is the 'System Matrix',

B is the 'Input Matrix',

C is the 'Output Matrix' and

D is the 'Feed forward Matrix'.

In order to do do the state space representation of the Mars UAV, the following states are selected from the system for a 6 DOF system. The states are denoted as described in Table (7.1).

State Name	State representation
Position along X axis	x
Position along Y axis	y
Altitude along Z axis	z
Velocity along X axis	$\dot{x}$
Velocity along Y axis	$\dot{y}$
Velocity along Z axis	$\dot{z}$
Roll angle	ϕ
Pitch angle	θ
Yaw angle	ψ
Roll rate	$\dot{\phi}$
Pitch rate	$\dot{\theta}$
Yaw rate	$\dot{\psi}$

Table 7.1: States representation

Therefore, the state vector can be represented as $x(t) = [x\ y\ z\ \dot{x}\ \dot{y}\ \dot{z}\ \phi\ \theta\ \psi\ \dot{\phi}\ \dot{\theta}\ \dot{\psi}]^T$. Also the Input matrix is the matrix that consists of required thrust and torque values. Therefore, Input vector can be represented as $u(t) = [U_1\ U_2\ U_3\ U_4]^T$. The Mars UAV system has 6 outputs from the modeling and the outputs are mainly linear and angular positions. Thus, the output matrix can be represented as $y(t) = [x \; y \; z \; \phi \; \theta \; \psi]^T$. For the purpose of state space representation, state vector $x(t) = X = [\, x_1 \; x_2 \; x_3 \; x_4 \; x_5 \; x_6 \; x_7 \; x_8 \; x_9 \; x_{10} \; x_{11} \; x_{12} \,]^T$. Therefore, the Mars UAV mathematical model can be written in state space representation as,

$$\dot{x}_1 = \dot{x} = x_4$$

$$\dot{x}_2 = \dot{y} = x_5$$

$$\dot{x}_3 = \dot{z} = x_6$$

$$\dot{x}_4 = \ddot{x} = [\sin\psi \sin\phi + \sin\psi \sin\theta \cos\phi] \frac{U_1}{m}$$

$$\dot{x}_5 = \ddot{y} = [-\cos\psi \sin\phi + \sin\psi \sin\theta \cos\phi] \frac{U_1}{m}$$

$$\dot{x}_6 = \ddot{z} = [\sin\theta \cos\phi] \frac{U_1}{m} - g$$

$$\dot{x}_7 = \dot{\phi} = x_7$$

$$\dot{x}_8 = \dot{\theta} = x_8$$

$$\dot{x}_9 = \dot{\psi} = x_9$$

$$\dot{x}_{10} = \ddot{\phi} = [\dot{\theta}\dot{\psi}(I_{yy} - I_{zz}) - J_r\dot{\theta}\Omega_r + U_2]\frac{1}{I_{xx}}$$

$$\dot{x}_{11} = \ddot{\theta} = [\dot{\phi}\dot{\psi}(I_{zz} - I_{xx}) - J_r\dot{\phi}\Omega_r + U_3]\frac{1}{I_{yy}}$$

$$\dot{x}_{12} = \ddot{\psi} = [\dot{\phi}\dot{\theta}(I_{xx} - I_{yy}) + U_4]\frac{1}{I_{zz}}$$

The equation (7.24) represents the overall state space representation of the Mars UAV mathematical model.

$$f(X,U) = \begin{bmatrix} x_4 \\ x_5 \\ x_6 \\ \ddot{x} = [\sin\psi\sin\phi + \sin\psi\sin\theta\cos\phi]\frac{U_1}{m} \\ \ddot{y} = [-\cos\psi\sin\phi + \sin\psi\sin\theta\cos\phi]\frac{U_1}{m} \\ [\sin\theta\cos\phi]\frac{U_1}{m} - g \\ x_7 \\ x_8 \\ x_9 \\ \ddot{\phi} = [\dot{\theta}\dot{\psi}(I_{yy} - I_{zz}) - J_r\dot{\theta}\Omega_r + U_2]\frac{1}{I_{xx}} \\ [\dot{\phi}\dot{\psi}(I_{zz} - I_{xx}) - J_r\dot{\phi}\Omega_r + U_3]\frac{1}{I_{yy}} \\ \ddot{\psi} = [\dot{\phi}\dot{\theta}(I_{xx} - I_{yy}) + U_4]\frac{1}{I_{zz}} \end{bmatrix} \tag{7.24}$$

Chapter 8

Mars UAV System Control

In this chapter the mathematical model of the Mars UAV system is used for open and closed loop simulations. The same model is also used for a controller design. With respect to the scope of this book project, The PID control is discussed in detail and a subsequent PID controller is developed in order to control Altitude, Position and Yaw of the Mars UAV vehicle. The controller gain are tuned manually in order to note the response of each rotational and transnational parameter and get the best possible fine tuned controller for trajectory tracking purpose.

8.1 Open Loop Simulation

Open loop simulations are not great at showcasing the system capabilities in terms of control. Although In order to verify the working of mathematical model of the Mars UAV, an open loop simulation was considered. The open loop simulation was done in Simulink. The Mars UAV parameters are not take into the mathematical model because the rotor size and rotor speed calculated for Mars operating conditions, do not give realistic results in Earth based model. Therefore in order to simplify the complexity of implementing original physical parameters of Mars UAV, a quadrotor parameters are assumed for mathematical model and for Matlab simulations. The quadrotor parameters considered for this purpose are from the PhD book of Samir Boubdallah [39]. The block diagram for open loop simulation of Mars UAV is shown in Figure 8.1. Initially the purpose of the open loop simulation was to give a set of input rotor speeds in order to make the system hover at certain altitude (Z). As shown in Figure 8.1, The set of input rotor speeds are given to the motor mixer of the system and from the motor mixer as an output, Thrust as well as roll, pitch and yaw moments are extracted. The roll pitch and yaw moments are fed into the rotational subsystem in order to compute for the angular positions. The detailed Simulink diagrams of rotational, translational subsystem and the motor mixer will be shown in coming sections.

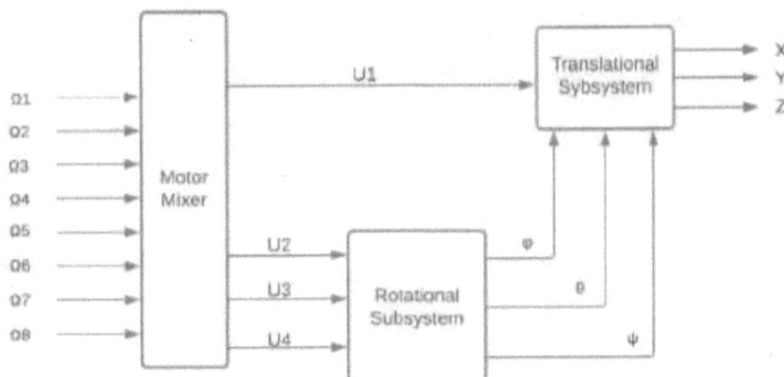

Figure 8.1: Block diagram of open loop simulation

In order to hover, the rotor speeds input can be calculated from the equation (8.1).

$$mg = 8F_T$$

$$mg = 8(K_T \Omega_i^2) \tag{8.1}$$

Where, Ω_i is the rotor speed for ith rotor for hover condition. In the hover condition it is assumed that the forces acting on the Mars UAV are the eight thrust vectors from the eight rotors acting upward and the gravitational force acting downwards. When the rotor speeds are given as input to the system it is noted and observed that all the state variables and their derivatives are zero. Only changing quantity is the Altitude of the system and is proportional to the thrust force produced. When a non zero positive value of rotor speed is fed into the system, the altitude change is observed. In case if the input rotor speeds are fed in such a way that the thrust is equal to the weight force, the vehicle hovers at constant altitude. By varying the eight rotor speed inputs, the roll pitch and yaw movement of the vehicle can also be observed. This observations state that the mathematical model made in Simulink is correct and this also gives foundation to build control system of the vehicle.

8.1.1   Motor Mixer Subsystem

For open loop simulation purpose, a specific simulink subsystem is designed in order to take the
rotor speeds as an input and process the input values in order to give the Thrust produced, roll,
pitch and yaw moments. The motor mixer is designed to take eight rotor speeds and a saturation
block is placed in order to limit the maximum and minimum rotor speeds feeding to the subsystem.

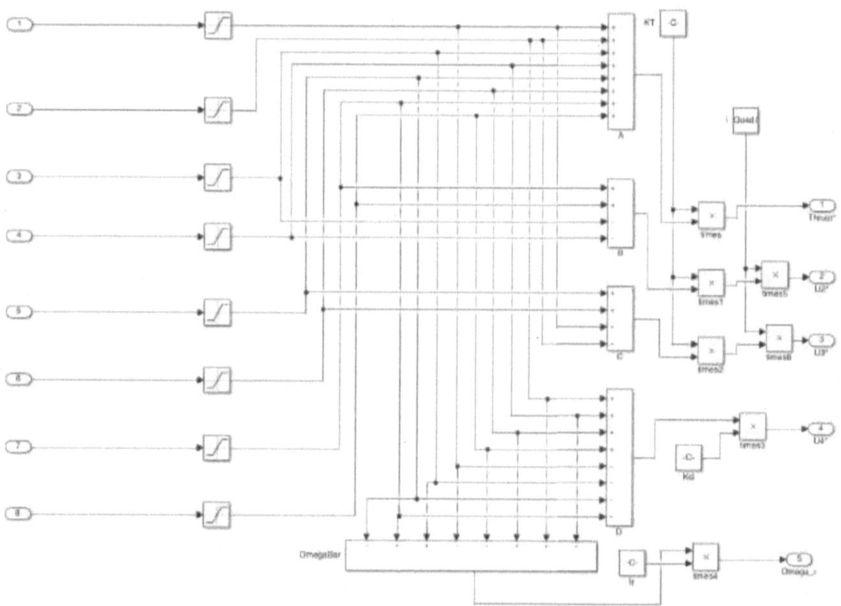

Figure 8.2: Motor mixer subsystem for open loop simulation

The inside of the motor mixer subsystem made in Simulink is shown in Figure 8.2. The 1 to 8
input blocks are the rotor speed values which are fed into summation blocks to compute for thrust.
The roll, pitch and yaw torque can be computed by coupling the inputs of different rotor speeds in
different manner as discussed in the definition of U_2, U_3 and U_4. The constants and coefficients are
loaded from a Matlab file which has the variables stored in it.

8.1.2 Rotation Subsystem

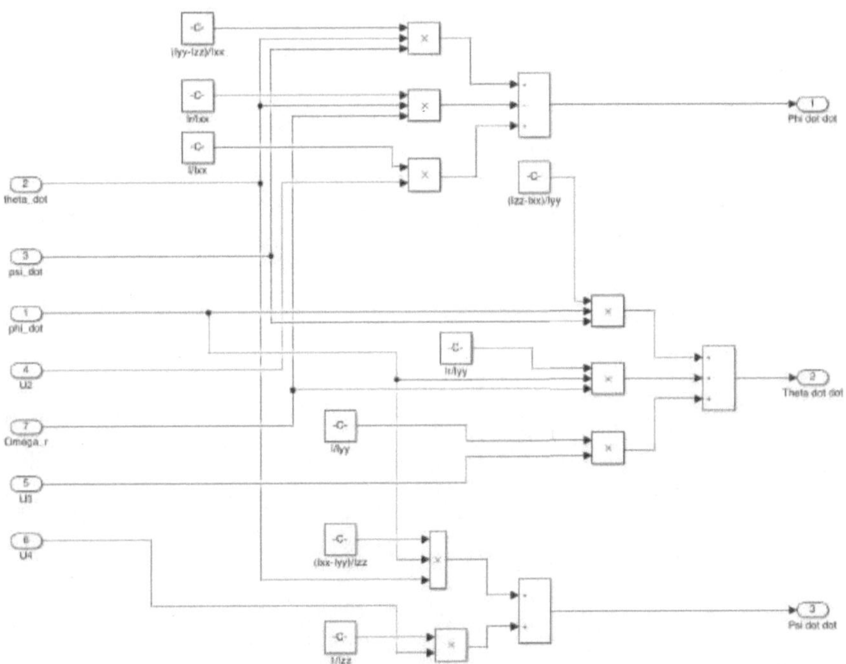

Figure 8.3: Rotation subsystem for open loop simulation

The Simulink block diagram of rotation subsystem for overall simulation is shown in Figure 8.3. For the purpose of open loop simulation, out of 6 inputs shown in Figure 8.3, only 3 inputs are necessary. Therefore, for a small angle approximation and considering all euler angles are close to zero, the angular rates are roughly equal to the time derivative of the euler angles. Therefore for simplification of modeling in open loop simulation, p, q and r are considered equal to $\dot{\phi}$, $\dot{\theta}$ and $\dot{\psi}$ respectively. With the help of transfer function, the angular acclerations in roll, pitch and yaw direction can be further integrated to feed the angular positions into the translational subsystem.

8.1.3　Translation Subsystem

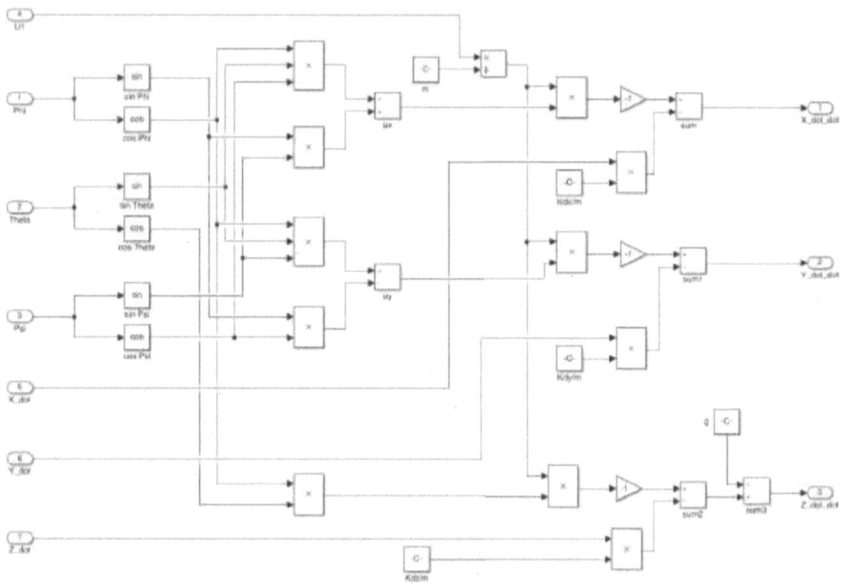

Figure 8.4: Translation subsystem for open loop simulation

The Simulink block diagram of the translation subsystem is shown in Figure 8.4. For the case of open loop simulation, the set of inputs fed into the translation subsystem are the angular positions ϕ, θ, ψ and the Thrust U_1. The Figure 8.4 shows the generic form of the translation subsystem because it also has $\dot{x}$, $\dot{y}$ and $\dot{z}$ as the set of inputs. however, these time derivatives of positions are integrated from the output of the translation subsystem and are then fed back into the subsystem. As an output from the translation subsystem, the accleration in x, y and z directions are received. These acclerations are integrated using the transfer function in order to compute for the position in Cartesian coordinates. The main output required for open loop simulation that focuses on hover mode is the altitude value.

8.2   Closed Loop Simulations

Once the open loop simulations were run in order to see the mathematical model of Mars UAV is correct, the model was then extended to make a closed loop control. The closed loop control includes Altitude, Attitude and position controller. The simulation frequency for closed loop control was set to 250 Hz and is within limits of a normal quadrotor system.[40]

8.2.1   Altitude Controller

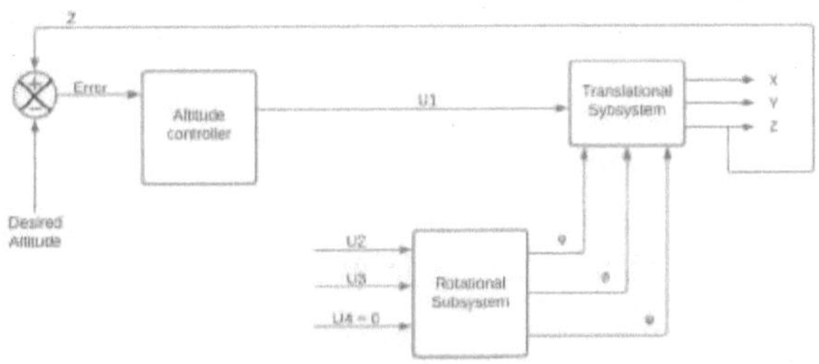

Figure 8.5: Block diagram of the Altitude controller

The altitude controller as shown in Figure 8.5 was implemented to the open loop model in order to get the actual and desired values of Z and feed the error into the system. The goal of the altitude controller is to take the error signal and process it to produce the Thrust or U_1 signal as an output to feed it into the translational subsystem.

8.2.2   Heading and Attitude Controller

The attitude and heading controller was implemented inside the open loop model of the Mars UAV in order to feel the heading angle as well as attitude of the system. The block diagram of the Heading and Attitude controller is shown in Figure 8.6. The working of the attitude controller is similar to the altitude controller, it takes the desired values of roll and pitch signal and get the error by subtracting it from the actual roll and pitch signal coming from the rotational subsystem. The attitude controller then feeds the error signal in order to control the attitude of Mars UAV.

The heading of the vehicle is defined as the ψ or yaw angle. So the heading controller takes the difference between desired heading and actual heading and feeds the error signal into the controller. [40]

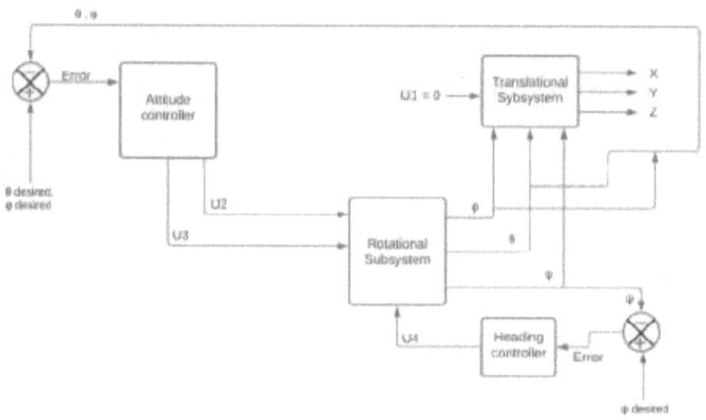

Figure 8.6: Block diagram of the Attitude and Heading controller

8.2.3 Position Controller

From the state space representation we know that the x and y position can not be directly controlled by the input matrix that consists of the $[U_1\ U_2\ U_3\ U_4]^T$. Therefore it is necessary to include a position controller for the Mars UAV system that consists of Altitude, Attitude and heading controller in order to adapt the position error and feed it into the controller. For mathematical modeling purpose, the x and y positions can be computed from the roll ϕ and pitch θ angles. From the equations of motion in translational direction we know that the linear accleration are given by,

$$\ddot{x} = [\sin\psi\sin\phi + \sin\psi\sin\theta\cos\phi]\frac{U_1}{m}$$

$$\ddot{y} = [-\cos\psi\sin\phi + \sin\psi\sin\theta\cos\phi]\frac{U_1}{m}$$

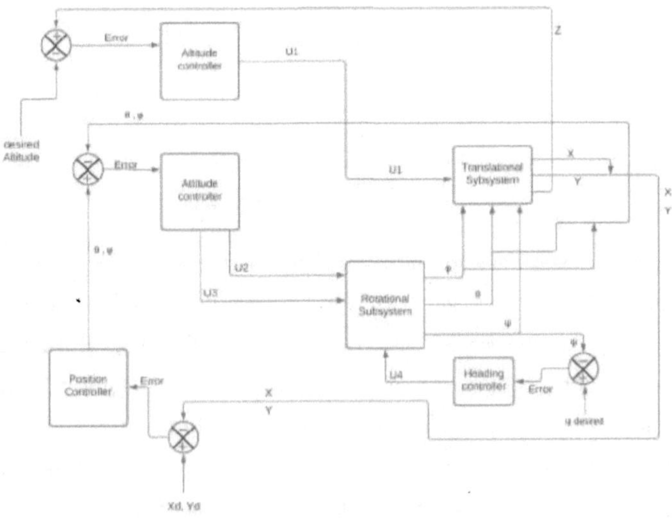

Figure 8.7: Block diagram of the Position controller

Because the hover condition is considered, it is allowed to take small angle implementation in roll and pitch directions. Therefore,

$$\ddot{x} = [\sin\psi\phi + \sin\psi\theta]\frac{U_1}{m}$$

$$\ddot{y} = [-\cos\psi\phi + \sin\psi\theta]\frac{U_1}{m}$$

And these equations can be written in matrix form to compute for roll and pitch angle as,

$$\frac{m}{U_1}\begin{bmatrix}\ddot{x}\\\ddot{y}\end{bmatrix} = \begin{bmatrix}\sin\psi & \cos\psi\\-\cos\psi & \sin\psi\end{bmatrix}\begin{bmatrix}\phi\\\theta\end{bmatrix}$$

In order to compute for the roll and pitch angles,

$$
\begin{bmatrix} \phi \\ \theta \end{bmatrix} = \begin{bmatrix} \sin\psi & \cos\psi \\ -\cos\psi & \sin\psi \end{bmatrix}^{-1} \frac{m}{U_1} \begin{bmatrix} \ddot{x} \\ \ddot{y} \end{bmatrix}
$$

$$
\begin{bmatrix} \phi \\ \theta \end{bmatrix} = \frac{m}{U_1} \begin{bmatrix} \ddot{x}\sin\psi - \ddot{y}\cos\psi \\ \ddot{x}\cos\psi + \ddot{y}\sin\psi \end{bmatrix} \tag{8.2}
$$

For this mathematical model to work under small angle approximation, for position controller the ϕ and θ values should lie in the range of -20° to 20°. From the equation (8.2) the angular positions of the Mars UAV can be computed and can be fed into the controller for position hold of the system. The block diagram of the position control of the system is shown in Figure 8.7. This controller includes all three, Altitude, Attitude and heading controller for the Mars UAV system. In this section basic block diagrams of the controllers are discussed, However the inside of the controller can include any kind of linear or nonlinear control algorithm. In all controllers the input signal is the error signal and the output signal can be a single or set of signals that could be U_1, U_2, U_3 or U_4.

8.3   PID Control

Once the Mars UAV mathematical model is developed in Matlab, a PID controller is developed. The PID controller takes the error signals and multiply them with the PID gains and feeds them to the Mars UAV system as an input. The block diagram of the PID controller is shown in Figure 8.8.

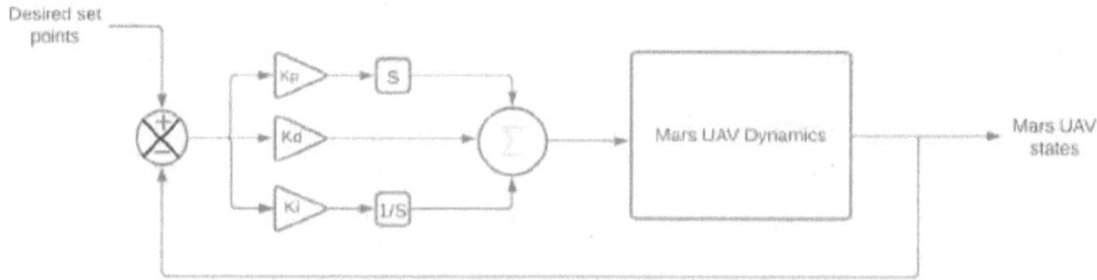

Figure 8.8: Block diagram of the PID controller

8.3.1 Altitude PID Control

The altitude PID controller basically generates the Thrust U_1 control signal in order to control the Z displacement or altitude of the system. The mathematical formulation of the altitude PID controller is based on equation (8.3).

$$U_1 = k_p(z_d - z) + k_d(\dot{z}_d - \dot{z}) + k_i \int (z_d - z)dt + mg \tag{8.3}$$

In equation (8.3),

k_p is Proportional gain,

k_d is Derivative gain,

k_i is Integral gain,

z_d is desired altitude,

$\dot{z}_d$ is desired rate of change of the altitude and

z is the altitude.

The simulink block diagram of the altitude PID controller is shown in Figure 8.9.

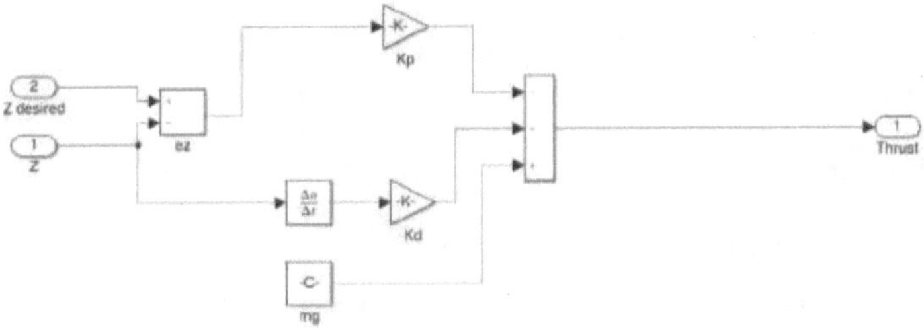

Figure 8.9: Block diagram of the altitude PID controller

8.3.2 Roll PID Control

In order to compute for the roll moment of U_2 signal, a roll PID controller is developed. The roll PID controller works on the basis of mathematical formulation shown in equation (8.4). This controller needs desired roll angle from the position controller and actual roll angle from the system states in order to calculate the roll signal U_2.

$$U_2 = k_p(\phi_d - \phi) + k_d(\dot{\phi}_d - \dot{\phi}) + k_i \int (\phi_d - \phi)dt \qquad (8.4)$$

In equation (8.4),

ϕ_d is desired roll angle,

$\dot{\phi}_d$ is desired rate of change of the roll and

ϕ is the roll angle from system states.

The simulink block diagram of the roll PID controller is shown in Figure 8.10.

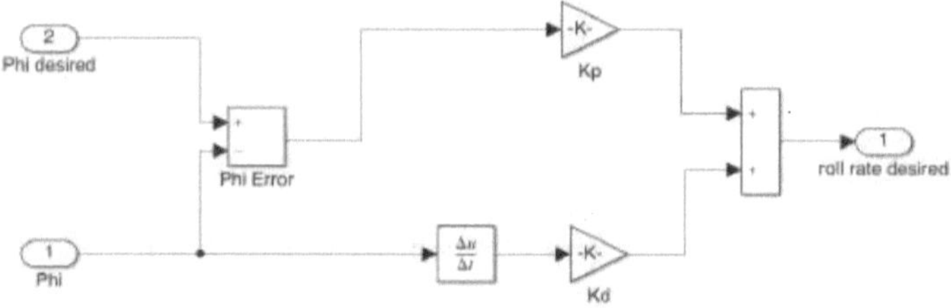

Figure 8.10: Block diagram of the Roll PID controller

8.3.3 Pitch PID Control

In order to calculate the pitch moment U_3, a pitch PID controller is designed. The pitch PID controller works on the basis of equation (8.5). The pitch controller takes desired pitch angle from the position controller and the actual pitch angle from the Mars UAV system states. The controller computes the error and multiplies the error signal with the gain in order to calculate the pitch moment U_3.

$$U_3 = k_p(\theta_d - \theta) + k_d(\dot{\theta}_d - \dot{\theta}) + k_i \int (\theta_d - \theta)dt \qquad (8.5)$$

In equation (8.5),

θ_d is desired pitch angle,

$\dot{\theta}_d$ is desired rate of change of the pitch and

θ is the pitch angle from system states.

The simulink block diagram of the pitch PID controller is shown in Figure 8.11.

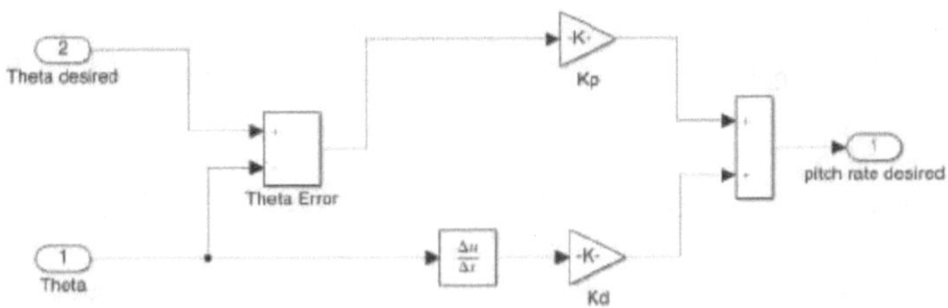

Figure 8.11: Block diagram of the Pitch PID controller

8.3.4 Yaw PID Control

In order to calculate the yaw moment U_4, a yaw PID controller is designed. The yaw PID controller is designed on the basis of equation (8.5). The yaw controller or the heading controller takes desired heading angle from the desired input and the actual heading angle from the system states. The controller computes the error and multiplies the error signal with the PID gains in order to calculate the yaw moment U_4.

$$U_4 = k_p(\psi_d - \psi) + k_d(\dot{\psi}_d - \psi) + k_i \int (\psi_d - \psi)dt \qquad (8.6)$$

In equation (8.6),

ψ_d is desired heading angle,

$\dot{\psi}_d$ is desired rate of change of the heading of the UAV and

ψ is the heading angle from system states.

The simulink block diagram of the yaw PID controller is shown in Figure 8.12.

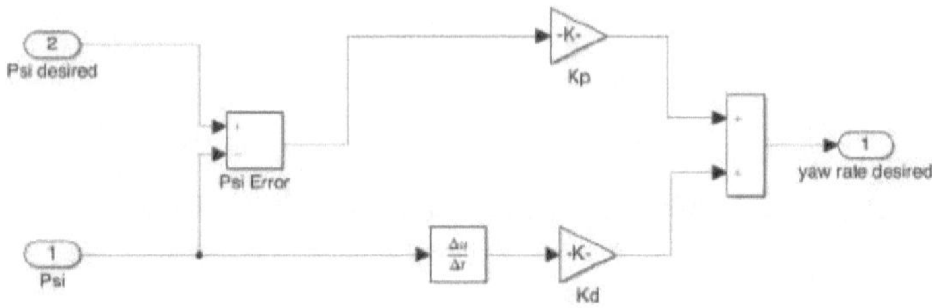

Figure 8.12: Block diagram of the Yaw PID controller

8.3.5 Position PID Control

Once the individual controllers for altitude, roll, pitch and heading are designed, a complete position controller can be designed in the same way the individual controllers were designed. We know as long as the small angle approximation holds, the linear accelerations are related to the roll and pitch angles from equation (8.2). Therefore,

$$\ddot{x} = k_p(x_d - x) + k_d(\dot{x}_d - \dot{x}) + k_i \int (x_d - x)dt \qquad (8.7)$$

$$\ddot{y} = k_p(y_d - y) + k_d(\dot{y}_d - \dot{y}) + k_i \int (z_y - y)dt \qquad (8.8)$$

In equations (8.7) and (8.8),

x_d is desired X position,

$\dot{x}_d$ is desired rate of change of the position in X,

x is the Position in X,

y_d is desired Y position,

$\dot{y}_d$ is desired rate of change of the position in Y direction and

y is the Position in y.

Once $\ddot{x}$ and $\ddot{x}$ are calculated from the equations (8.7) and (8.8), the values can be fed into the equation (8.2) in order to calculate the roll and pitch angle. These ϕ and θ are then fed into the roll and pitch controller as discussed in the individual controller design in order to compute the roll and pitch moments. For simplification of all control simulations, only PD control is implemented in Matlab.

8.4   Simulation Results for PID Control

As discussed in the previous section on the design of individual PID controllers, a set of input are given in order to see the response from the PID controller. The gains of all controllers are tuned manually. The manual tuning of the gains has led to observing the response from PID controller to get as close as possible to the desired input. In order to see the response of altitude controller, a desired input of 4 m is supplied to the system. The response of the altitude controller is shown in Figure 8.13. Steady state error is not recorded. The PID controller is simplified to a PD controller in order to minimize the steady state error.

P gain is set to 3,

D gain is set to -2.5,

Desired Altitude is 4m,

Observed overshoot is 1.04%.

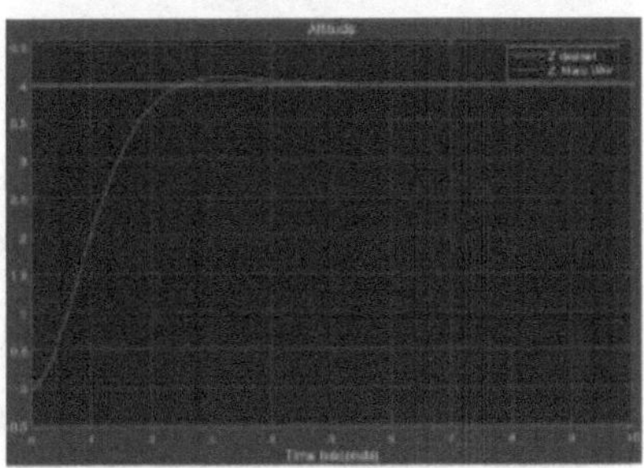

(a) Altitude response

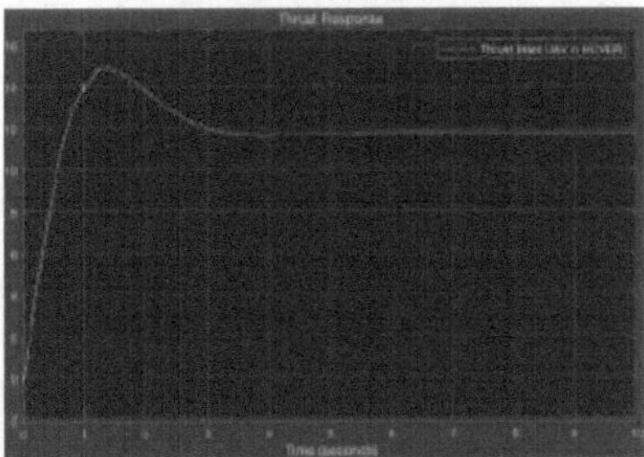

(b) Hover thrust response for Z = 4 m.

Figure 8.13: Altitude response PID control

8.4.1   Position response

In order to see the control responses from the position PID controller a set of x and y position input
was given to the system. The input, and gains are described in table 8.1.

Position	Desired Input	P gain	I gain	D gain
X	1	1.2	0.05	-0.4
Y	1	1.2	0.05	-0.4
Z	0	3	0	-2.5

Table 8.1: PID position inputs and gains

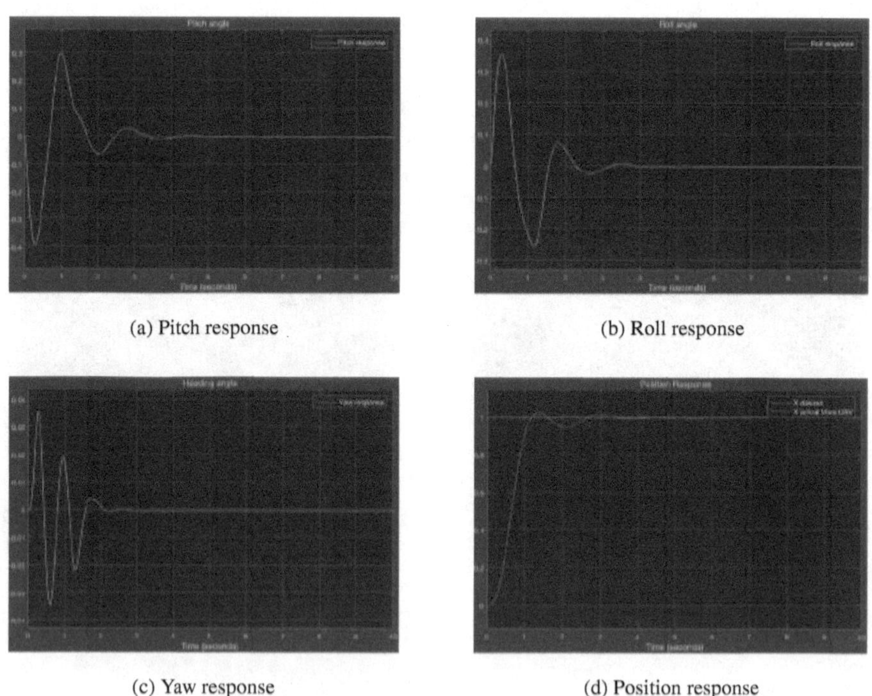

(a) Pitch response   (b) Roll response

(c) Yaw response   (d) Position response

Figure 8.14: Control responses from PID controller

From Figure 8.14, it can be observed that the PID controller minimizes the error signal and feeds it to the respective subsystem. The control responses observed in Figure 8.14 make it clear about the roll, pitch and yaw signal the Mars UAV model generates in order to reach the desired position in XY plane. The position response overshoots by a small margin but eventually converges to the desired input.

8.4.2   Trajectory Follow response

Once the gains were tuned for position control of the Mars UAV PID controller, the controller input signals were then extended to see if the model follows a desired trajectory. The first part of the trajectory follow response is discussed for a 2D trajectory in XY plane. In quadrotor system modelling, a square trajectory is the perfect example to observe the control responses of the system Because the square trajectory has very sharp 90°turns, the control response of the system can be observed at it's extreme limits.

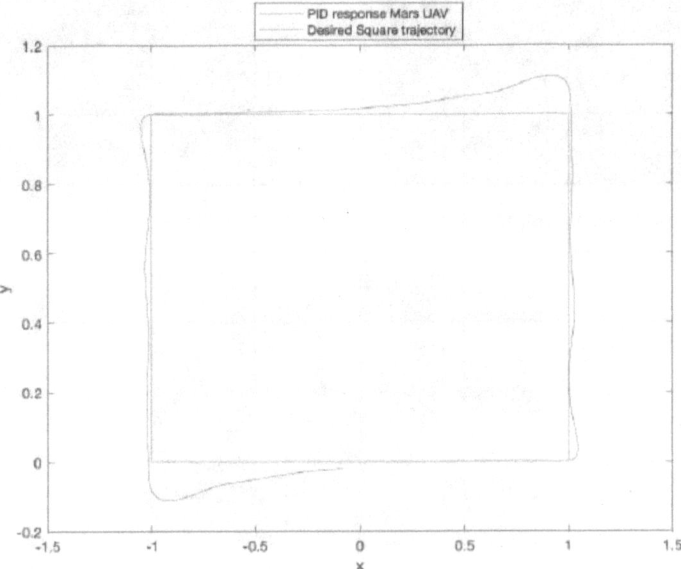

Figure 8.15: Square trajectory follow response

The PID response of the Mars UAV while following a square trajectory is shown in Figure 8.15. As seen from Figure 8.16 (a) the peaks of input signal for X position is changing twice therefore, the overshoot in X direction is not large as compared to the Y overshoot. The main reason for a significant overshoot in y direction at the diagonal corners of the square is the time duration spent for the UAV in Y direction of motion is more from the input signal. Therefore, the UAV travels in Y direction at a higher velocity because there is no intermediate signal to slow down the vehicle. The PID controller can be tuned fine in order to minimize the overshoot. However in order to see the working of PID control concept and vehicle response for a square trajectory, the overshoot is highlighted in Y direction. For square trajectory following the output roll signal is also shown in Figure 8.16 (c). It can be observed that the roll signal is given to the motor mixer in order to match the input and output states with a PID control.

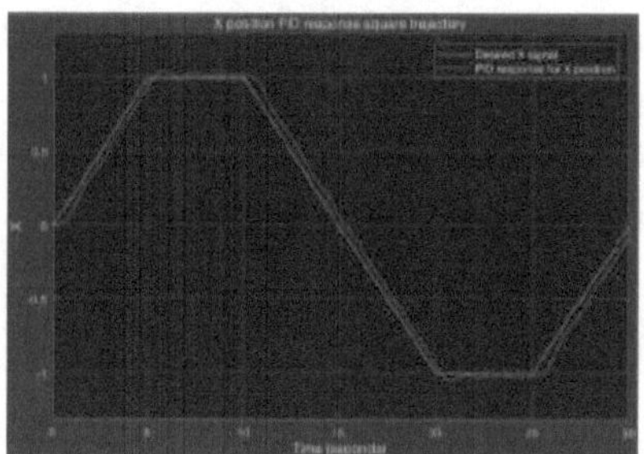

(a) X position response

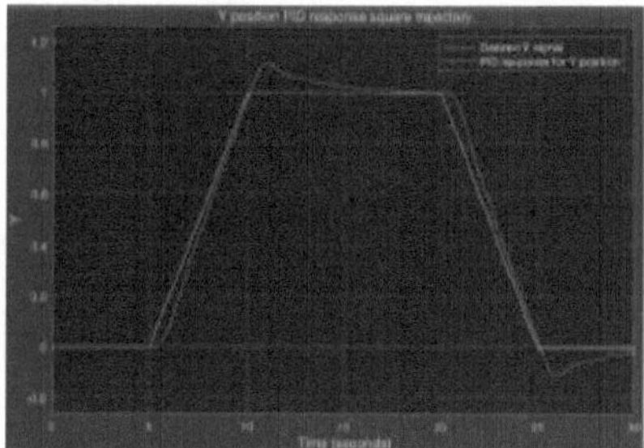

(b) Y position response

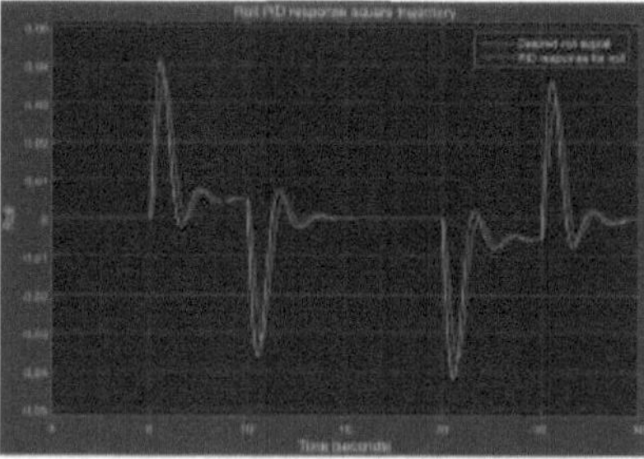

(c) Roll response

Figure 8.16: Control responses for square trajectory follow with PID control

The next part of the trajectory follow response is discussed for a 3D trajectory which has inputs in X, Y as well as Z direction. A good example of a 3D trajectory is a helix trajectory for a UAV system to follow. In order give inputs for the helix trajectory, a Sine and Cos inputs are selected. The Z input is based on the simulation time.

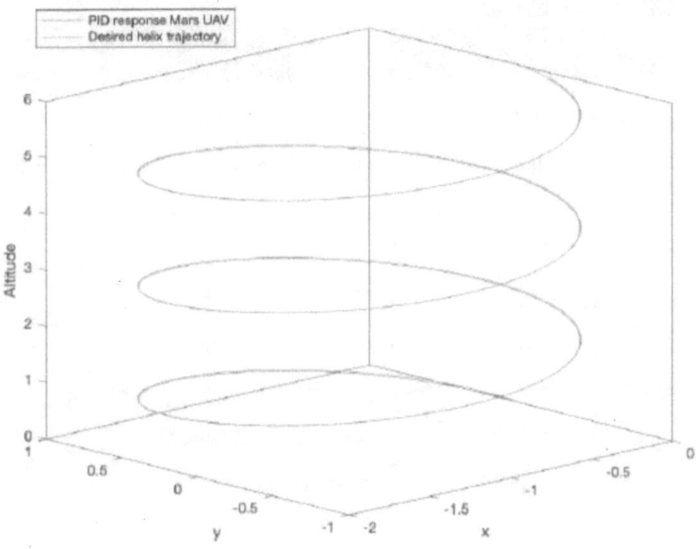

Figure 8.17: Helix trajectory follow response

The Figure 8.17 shows the helix trajectory follow response of the Mars UAV. After understanding the effect of gains on response, the PID gains were tuned finely in order to match the input helix trajectory. Because the start position of the Mars UAV is at origin, the circle of the helix is shifted to (X Y) = (-1 0). This is done by giving a bias to the X sine wave in the input block of the helix trajectory in Simulink. The thrust, roll and pitch responses of the Mars UAV in order to follow the helix trajectory are shown in Figure 8.18. In order to conclude the PID control simulation section, it can be stated that the PID controller are good at suppressing the error signal at basic level by manually tuning the PID gains. In order to advance the response of the PID controller, several linear and non linear algorithms can be adapted in order to tune the gains of the PID controller.

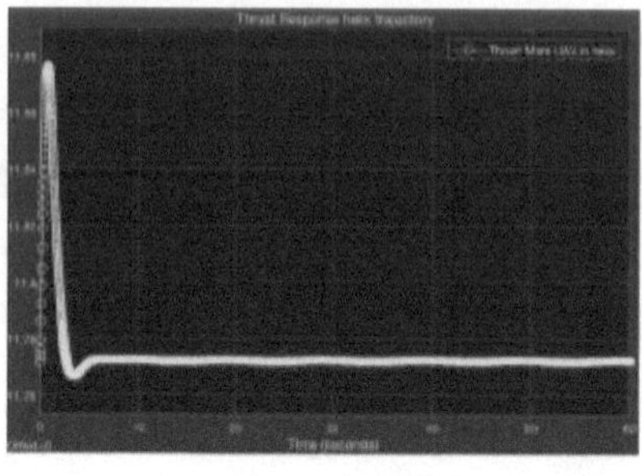

(a) Thrust response

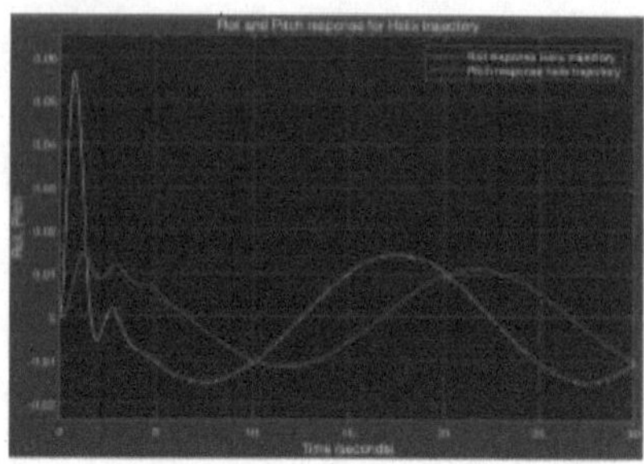

(b) Roll and Pitch response

Figure 8.18: Control responses for helix trajectory follow with PID control

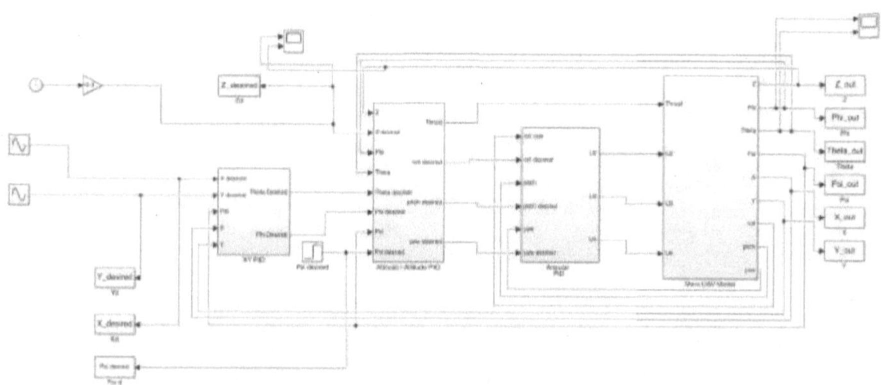

Figure 8.19: Simulink Model of Mars UAV

Chapter 9

Conclusion

The goal of this master book project was to do flow simulations for a flight of UAV in thin atmosphere of Mars and also to develop a mathematical model of the system in order to implement control strategies. Because surface pressure is very low and the atmosphere density of Mars is also low as compared to earth, the conventional rotor crafts are not able to fly in Martian conditions. In order to design a vehicle that has aerodynamic capability to generate lift in thin atmosphere, a multi rotor UAV is considered in many studies. This book project summarise the studies done for 2D airfoil optimization for low reynolds number flow. The 2D analysis have been carried out for flat plate, cambered plate as well as cambered airfoils. The studies suggest that in low reynolds number flow the main issue that occurs is flow transits into turbulent from laminar quite quickly at the leading edge. In order to delay the transition, the optimal flow velocity is calculated for each plate or airfoil. The flow also shows characteristics of turning supersonic towards the tip of the rotor blade. The low speed of sound in Mars atmosphere is problem and it generates shock waves which further lower the lift producing capability of the rover. In order to optimize the RPM speed for the set of co axial rotor, the 3D flow simulations were done in ANSYS Fluent. The simulation show that having a co axial system of rotors help in increasing the overall thrust of the system while also lowering the RPM at which each rotor needs to spin. The velocity and pressure contours strongly support a system of co axial rotors to be used for a UAV to fly in Mars atmosphere. The pressure force over the surface of the rotors is integrated to extract the thrust force produced by the rotor system.

For the Control part of the Mars UAV system, a mathematical model is designed in Matlab. The mathematical model differs from conventional quadrotor in motor mixer design because the Mars UAV uses in total eight rotors which are mounted in pairs in co axial configuration. Open loop simulations were carried out to test the correct working of the mathematical model of the Mars UAV. In order to control the Mars UAV mode for hover position or go to a pose condition, a PID controller is developed. The PID controller is simplified to a PD controller for most of the subsystems of the Mars UAV. The control responses of each signal are recorded and then analysed for gain tuning in order to minimize the error. The gains are tuned manually for this book project. However as a future work an algorithm can be implemented to tune the gains of the system. The PID controller has been a basic foundation to control the Mars UAV model in simulations and also to make it semi autonomous. As future work, in order to make the system fully autonomous a Linear Quadratic Regulator control and Model Predictive Control can be implemented into the system. The PID controller is a basic way to control the system for hovering, trajectory following purpose etc. Whereas, the LQR and MPC can really advance the system to make it more accurate. As a learning takeaway from this book project, the understanding of UAV dynamics and control has inspired to do detail study in making mathematical model of multi rotors for different purpose.

References

[1]   B. T. Pipenberg, M. Keennon, J. Tyler, B. Hibbs, S. Langberg, J. Balaram, H. F. Grip and J. Pempejian, 'Design and fabrication of the mars helicopter rotor, airframe, and landing gear systems,' in *AIAA Scitech 2019 Forum*, 2019, p. 0620 (cit. on pp. 2, 4, 22).

[2]   M. Johansson, 'Experimental and computational evaluation of capabilities of predicting aerodynamic performance for a mars helicopter rotor,' Ph.D. book, MS book, Department of Applied Mechanics, Chalmers University of . . ., 2017 (cit. on pp. 3, 8).

[3]   M. Bangura, M. Melega, R. Naldi and R. Mahony, 'Aerodynamics of rotor blades for quadrotors,' *arXiv preprint arXiv:1601.00733*, 2016 (cit. on pp. 8–10, 12).

[4]   M. K. Rwigema, 'Propeller blade element momentum theory with vortex wake deflection,' in *27th International congress of the aeronautical sciences*, vol. 2010, 2010, pp. 2–3 (cit. on p. 12).

[5]   W. J. Koning, E. A. Romander and W. Johnson, 'Performance optimization of plate airfoils for martian rotor applications using a genetic algorithm,' 2019 (cit. on pp. 16, 17).

[6]   W. J. Koning, E. A. Romander and W. Johnson, 'Low reynolds number airfoil evaluation for the mars helicopter rotor,' 2018 (cit. on pp. 16, 17, 20, 21, 23–25, 27–30).

[7]   F. W. Schmitz, *Aerodynamics of the model airplane*. Translation Branch, Redstone Scientific Information Center, Research and . . ., 1970 (cit. on pp. 17, 19, 20, 26).

[8]   A. Rezende and A. Nieckele, 'Evaluation of turbulence models to predict the edge separation bubble over a thin aerofoil,' in *Proceedings of the 20 th International Congress of Mechanical Engineering–COBEM*, 2009 (cit. on p. 17).

[9]   E. Laitone, 'Wind tunnel tests of wings at reynolds numbers below 70 000,' *Experiments in fluids*, vol. 23, no. 5, pp. 405–409, 1997 (cit. on p. 17).

[10]   H. Werlé *et al.*, 'Le tunnel hydrodynamique au service de la recherche aérospatiale.,' 1974 (cit. on pp. 17, 19).

[11]   M. Van Dyke and M. Van Dyke, 'An album of fluid motion,' 1982 (cit. on pp. 17, 19, 26).

[12]   A. V. Boiko, G. R. Grek, A. V. Dovgal and V. V. Kozlov, *The origin of turbulence in near-wall flows*. Springer Science & Business Media, 2013 (cit. on p. 17).

[13]   F. H. Sighard, 'Fluid dynamic drag: Practical information on aerodynamic drag and hydrodynamic resistance,' *Sighard, Hoerner F*, 1965 (cit. on pp. 18, 19).

[14] W. J. Koning, E. A. Romander and W. Johnson, 'Low reynolds number airfoil evaluation for the mars helicopter rotor,' 2018 (cit. on pp. 18, 19).

[15] M. J. Crompton, 'The thin aerofoil leading edge separation bubble,' Ph.D. book, Uni-versity of Bristol, 2001 (cit. on p. 19).

[16] E. Laitone, 'Wind tunnel tests of wings at reynolds numbers below 70 000,' *Experiments in fluids*, vol. 23, no. 5, pp. 405–409, 1997 (cit. on pp. 19, 25, 28).

[17] L. E. B. Sampaio, A. L. T. Rezende and A. O. Nieckele, 'The challenging case of the tur-bulent flow around a thin plate wind deflector, and its numerical prediction by les and rans models,' *Journal of Wind Engineering and Industrial Aerodynamics*, vol. 133, pp. 52–64, 2014 (cit. on p. 20).

[18] M. Anyoji, D. Numata, H. Nagai and K. Asai, 'Effects of mach number and specific heat ratio on low-reynolds-number airfoil flows,' *AIAA journal*, vol. 53, no. 6, pp. 1640–1654, 2015 (cit. on pp. 20, 25, 26).

[19] W. J. Koning, W. Johnson and B. G. Allan, 'Generation of mars helicopter rotor model for comprehensive analyses,' 2018 (cit. on pp. 23–25).

[20] W. Koning, 'Mars helicopter rotor aerodynamics and modeling,' NASA CR-2018-219735, Moffett Field, California, Tech. Rep., 2018 (cit. on p. 23).

[21] P. Lissaman, 'Low-reynolds-number airfoils,' *Annual review of fluid mechanics*, vol. 15, no. 1, pp. 223–239, 1983 (cit. on p. 23).

[22] T. J. Mueller and J. D. DeLaurier, 'Aerodynamics of small vehicles,' *Annual review of fluid mechanics*, vol. 35, no. 1, pp. 89–111, 2003 (cit. on p. 23).

[23] B. Carmichael, 'Low reynolds number airfoil survey, volume 1,' 1981 (cit. on p. 23).

[24] W. J. Koning, W. Johnson and H. F. Grip, 'Improved mars helicopter aerodynamic rotor model for comprehensive analyses,' *AIAA Journal*, vol. 57, no. 9, pp. 3969–3979, 2019 (cit. on pp. 25, 32, 33).

[25] R. H. Nichols and P. G. Buning, 'User's manual for overflow 2.1,' *University of Alabama at Birmingham, Birmingham, AL*, 2008 (cit. on p. 25).

[26] P. J. Kunz and I. Kroo, 'Analysis and design of airfoils for use at ultra-low reynolds num-bers,' *Fixed and flapping wing aerodynamics for micro air vehicle applications*, vol. 195, pp. 35–59, 2001 (cit. on p. 26).

[27] H. Werle, 'Le tunnel hydrodynamique au service de la recherche aérospatiale, publication no 156,' *ONERA, Oficce National d'études et de recherches aérospaciales*, 1974 (cit. on p. 26).

[28] R. Radespiel, J. Windte and U. Scholz, 'Numerical and experimental flow analysis of moving airfoils with laminar separation bubbles,' *AIAA journal*, vol. 45, no. 6, pp. 1346–1356, 2007 (cit. on p. 26).

[29] J. Winslow, H. Otsuka, B. Govindarajan and I. Chopra, 'Basic understanding of airfoil characteristics at low reynolds numbers (10 4–10 5),' *Journal of Aircraft*, vol. 55, no. 3, pp. 1050–1061, 2018 (cit. on p. 26).

[30] S. Sunada, A. Sakaguchi and K. Kawachi, 'Airfoil section characteristics at a low reynolds number,' 1997 (cit. on p. 28).

[31] S. Sunada, T. Yasuda, K. Yasuda and K. Kawachi, 'Comparison of wing characteristics at an ultralow reynolds number,' *Journal of aircraft*, vol. 39, no. 2, pp. 331–338, 2002 (cit. on p. 28).

[32] M. Okamoto, K. Yasuda and A. Azuma, 'Aerodynamic characteristics of the wings and body of a dragonfly,' *Journal of experimental biology*, vol. 199, no. 2, pp. 281–294, 1996 (cit. on p. 28).

[33] J. Balaram, I. Daubar, J. Bapst and T. Tzanetos, 'Helicopters on mars: Compelling science of extreme terrains enabled by an aerial platform,' *LPICo*, vol. 2089, p. 6277, 2019 (cit. on pp. 33, 34).

[34] B. Rubi, B. Morcegol and R. Peréz, 'Adaptive nonlinear guidance law using neural networks applied to a quadrotor,' in *2019 IEEE 15th International Conference on Control and Automation (ICCA)*, IEEE, 2019, pp. 1626–1631 (cit. on p. 48).

[35] S. R. B. dos Santos, C. L. Nascimento and S. N. Givigi, 'Design of attitude and path tracking controllers for quad-rotor robots using reinforcement learning,' in *2012 IEEE Aerospace Conference*, IEEE, 2012, pp. 1–16 (cit. on p. 49).

[36] F. Sabatino, *Quadrotor control: Modeling, nonlinearcontrol design, and simulation*, 2015 (cit. on pp. 49, 50, 52).

[37] G. B. Raharja, K. G. Beom and Y. Kwangjoon, 'Design and implementation of coaxial quadrotor for an autonomous outdoor flight,' in *2011 8th International Conference on Ubiquitous Robots and Ambient Intelligence (URAI)*, IEEE, 2011, pp. 61–63 (cit. on p. 53).

[38] Z. Tahir, W. Tahir and S. A. Liaqat, 'State space system modeling of a quad copter uav,' *arXiv preprint arXiv:1908.07401*, 2019 (cit. on p. 56).

[39] S. Bouabdallah, 'Design and control of quadrotors with application to autonomous flying,' Epfl, Tech. Rep., 2007 (cit. on p. 60).

[40] M. K. Habib, W. G. A. Abdelaal, M. S. Saad *et al.*, 'Dynamic modeling and control of a quadrotor using linear and nonlinear approaches,' 2014 (cit. on pp. 65, 66).

www.ingramcontent.com/pod-product-compliance
Lightning Source LLC
LaVergne TN
LVHW091618170726
843492LV00007B/2475